THE MULTINATIONAL CONSTRUCTION INDUSTRY

Howard Seymour, formerly Department of Economics, University of Reading and currently Construction Analyst at Kitcat and Aitken Stockbrokers.

Over the past ten years competition in the international construction industry has increased dramatically as new contractors have entered the industry from less developed countries and as workloads have dropped in the major markets. In the changed environment international contractors face a complex situation in which success only partially depends on price competition; and other factors which differentiate a contractor's product from that of competitors become more crucial. This book, based on extensive original research, outlines the motives and methods of overseas operations by international contractors. Drawing on an economic analysis of the industry and on elements of international investment and production theory the book discusses the problems of both individual enterprises and the major nationality groups in the industry (UK, USA, Japan, South Korea, France, Germany, Italy and Turkey). It surveys the major competitive features of the industry; contractors' organisational hierarchies; and the markets in which firms operate. It argues that project financing arrangements and home government support for contractors are major determinants of success; and it recommends that contractors reassess their overseas operations as part of a globally co-ordinated strategy if they are to remain competitive in the next five to ten years.

The Multinational Construction Industry

♦

Howard Seymour

CROOM HELM
London • New York • Sydney

© 1987 Howard Seymour
Croom Helm Ltd, Provident House, Burrell Row,
Beckenham, Kent, BR3 1AT

Croom Helm Australia, 44-50 Waterloo Road,
North Ryde, 2113, New South Wales

Published in the USA by
Croom Helm
in association with Methuen, Inc.
29 West 35th Street,
New York, NY 10001

British Library Cataloguing in Publication Data

Seymour, Howard
  The multinational construction industry.
  1. Construction industry 2. International
  business enterprises
  I. Title
  338.8'87   HD9715.A2
  ISBN 0-7099-5438-7

Library of Congress Cataloging-in-Publication Data

Seymour, Howard, 1961-
  The multinational construction industry / Howard Seymour.
    p.   cm.
  Bibliography: p.
  Includes index.
  1. Construction industry. 2. International business enterprises.
I. Title.
HD9715.A2S49   1987   87-21720
338.8'87 — dc 19
ISBN 0-7099-5438-7

Printed and bound in Great Britain by Mackays of Chatham Ltd, Kent

CONTENTS

1. Chapter 1: Introduction ... 1

   1.1 The international construction environment ... 1
   1.2 Developments in analysis of the industry ... 7
   1.3 Definition of term ... 9
   1.4 The research framework ... 10
   1.5 Objectives and hypotheses of the research ... 11
   1.6 Footnotes ... 14

2. Chapter 2: A Literature Review of Multinational Enterprise Theory ... 16

   2.1 Behavioural aspects of foreign investment ... 19
   2.2 Industrial Organisation Theory ... 20
   2.3 Internalisation theory and the MNE ... 31
   2.4 Location theory and the MNE ... 48
   2.5 General theories of the MNE ... 54

3. Chapter 3: An Economic Overview of the Construction Industry ... 58

   3.1 The client in construction ... 59
   3.2 Raw materials ... 61
   3.3 The construction process ... 62

Contents

3.4	The final product	68
3.5	Demand and Supply in the construction industry	70
3.6	Structure of the construction industry	74
3.7	Aspects of international construction	79

4. Chapter 4: Application of Multinational Enterprise Theory to International Construction — 85

4.1	The research framework	85
4.2	Ownership advantages	87
4.3	Location advantages	100
4.4	Internalisation advantages	106

5. Chapter 5: Empirical Analysis of Ownership Advantages — 127

5.1	Methodology of the research	127
5.2	Ownership advantages	129
5.3	Firm specific factors	129
5.4	Country specific advantages	143
5.5	Footnotes	176

6. Chapter 6: Internalisation and Locational Factors — 177

6.1	Internalisation factors	177
6.2	Locational advantages	196
6.3	Political risk in international contracting	225
6.4	Oligopolistic reaction in international construction	229
6.5	Footnotes	230

7. Chapter 7: The Financing of International Projects — 232

7.1	Financing operations in the MNE	233
7.2	The export credit mechanism	237

Contents

	7.3	Export credit financing as a country specific 0 advantage	240
	7.4	Summary	257
	7.5	Footnotes	258
8.		Chapter 8: Summary and Conclusions	260
	8.1	Summary of the thesis	260
	8.2	Relevance of the hypotheses of the research	262
	8.3	Theoretical conclusions	263
	8.4	Practical conclusions	265
	8.5	Limitations of the research and suggestions for future work	272
	8.6	Footnotes	274
9.		References	276
10.		Bibliography	287
		Subject Index	290
		Author Index	294

LIST OF TABLES

3.1	Client demands for construction works	61
3.2	Assessment of the level of competition in the construction industry	80
4.1	Country specific factors that generate ownership advantages	94
4.2	Country specific factors that generate location advantages	102
4.3	Strategic issues affecting the choice of contractual arrangements in international construction	114
4.4	Options open to the international contractor in foreign markets	118
5.1	Response rate of contractors approached	128
5.2	Sample breakdown of participants in survey	128
5.3	Perceived competitive advantages/disadvantages of firms with respect to Arab contractors	130
5.4	Perceived competitive advantages/disadvantages of firms with respect to other international contractors	132
5.5	Services offered by firms in survey	134
5.6	Services demanded by Middle Eastern clients	136
5.7	Types of bonds required within international construction	142
5.8	Size of top 10 contractors according to total value of awards 1982	148
5.9	Number of South Korean nationals working in the Middle East 1975-78	150

List of Tables

5.10	Top 25 international contractors in general building construction 1981	150
5.11	Top 25 international contractors in power and process plant construction 1981	151
5.12	Diversity of services offered by top 15 contractors from major contracting countries 1982	152
5.13	Top 20 lead managers in syndicated loans 1981	157
5.14	Top 200 international design consultants in developing countries by nationality 1983	161
5.15	Defence and capital goods sales in Middle East by nationality of seller 1984 (first half)	163
5.16	Home government legislation of direct influence on contractors' overseas operations	164
5.17	Summary chart of major contractors' country specific advantages	170
6.1	Participants reasons for not undertaking licensing of the firm's name	178
6.2	The hierarchy of control of the international construction firm	180
6.3	Participants' reasons for opening local subsidiary or changing from exporting to establishing a local office/subsidiary	186
6.4	Relationship between HQ and Middle Eastern subsidiary of participants in survey	188
6.5	Advantages and disadvantages of joint venturing	191
6.6	Initial incentives to overseas work of participants in survey	198
6.7	Locational influences on contractors in Middle Eastern market	200
6.8	Major incentives of location by mean	203
6.9	Major disincentives of location by mean	204
6.10	Factors considered important as criteria of temporary/permanent subsidiary set up from survey results	205
6.11	Share of Middle Eastern market by country of origin of contractor 1980-84	212
6.12	Share of South East Asian market by country of origin of contractor 1980-84	216
6.13	Share of African market by country of origin of contractor 1980-84	220

List of Tables

6.14	Share of Latin American market by country of origin of contractor 1980-84	222
6.15	Political risk measures taken by contractors in survey	228
7.1	A comparison of export financing systems of major contracting countries	242
7.2	Percentage cover of commercial and political risks of major contracting countries	245
7.3	OECD arrangement terms on officially supported export credits 1985	246
7.4	Effective interest rate with aid component of export finance	249
7.5	Aid commitment of major contracting countries 1983	253
7.6	Tying status of bilateral aid provided by major contracting countries 1983	255

LIST OF FIGURES

1.1	Invisible earnings of UK construction industry overseas 1978-83	4
1.2	Invisible earnings of UK construction industry overseas 1978-83, real changes	5
1.3	Foreign earnings by nationality of top 250 international contractors 1984	6
3.1	The construction process	58
3.2	Construction as a vertical relationship	65
4.1	Buckley and Casson model for a simple product	109
4.2	Buckley and Casson model for a complex product	110
5.1	Overseas awards of major contracting countries 1980-84	144
5.2	Domestic awards of major contracting countries 1980-84	147
5.3	UK consultants' overseas earnings 1974-84	159
6.1	Subsidiary operation of the international contractor	182
6.2	Pure export or 'one off' operation of the international contractor	183
6.3	Export and FDI situation where local office has construction capabilities (ie LO/PO)	184
6.4	Export/FDI situation with separate LO and PO facilities	185
6.5	Location of offices/subsidiaries and construction works carried out by survey participants 1982	206

ABSTRACT

Over the past ten years the international construction industry has become characterised by fierce competition as contractors have entered the industry from the less developed regions of the world in search of hard currency, and workloads have dropped in the major regional markets. As a consequence the environment in the industry has changed so that the international contractor in the 1980s faces a complex competitive situation in which success only partially depends upon price competition, and increasingly reflects the need for other factors that differentiates the contractor's product from all others.

By integrating an economic assessment of the international construction industry with elements of the theoretical framework of international investment and production the objective of the research outlined in this thesis is to outline and justify the motives and methods of overseas operations of the international contractor. This has been carried out for both the individual contracting enterprise and the major nationality groups in the industry (UK, USA, Japan, South Korea, France, Germany, Italy and Turkey). Much of the empirical analysis comes from a questionnaire survey carried out on twenty major international construction companies with offices in the UK, although also relies on government publications of the respective countries and information from trade magazines for the macro analysis of markets and competitors in the industry. The combination of the theoretical predictions and empirical study provide interesting results both for economic theory and for the contractor working in the industry, that involves outlining the major competitive

Abstract

features of the industry in relation to the contractor's organisational hierarchy and the markets in which firms operate. Above all the analysis tends to suggest that the theoretical framework is a useful tool for investigation of the industry, and that project financing arrangements and home government support for contractors are major determinants of success in the industry. The thesis recommends that contractors need to assess their overseas interests as part of a globally coordinated strategy if they are to remain competitive in the next 5-10 years.

ACKNOWLEDGEMENTS

This thesis and the research I have been involved with would not have been possible without the cooperation and help of many people. I am therefore grateful to all who have contributed to my work over the past three years and hope that it accurately reflects their input. All errors remain my responsibility alone.

The empirical work of the thesis owes a great deal to all those people involved with the international construction industry who were willing to discuss the subject with me. Confidentiality prevents the mentioning of individuals names, but I would like to thank all the executives of the companies which I visited, and also the consulting engineers, bankers, civil servants and journalists for their time.

For much help and advice throughout my research I am deeply indebted to Professor J.H. Dunning and Professor Roger Flanagan, who have provided continued support and constructive criticism through my years at Reading. Professor George Norman also deserves thanks for his encouragement in the first and subsequent years of my research, and for introducing me to the topic of international construction and showing me the high standards expected in a PhD. Several of the staff and postgraduates of the Economics department at Reading are thanked for helping to clarify certain matters and discussing various aspects of the research, and Mrs J. Turner and Mrs M. Lewis are acknowledged particularly for help with typing relating to my research. Mr P. Hodgson of the Major Projects Association, Templeton College Oxford, introduced me to the complex world of project finance and export credit funding and I would like to thank him in this respect.

Acknowledgements

Personal thanks go to many of the above, but also include various friends from Workington, Maryport, Liverpool and Reading who provided relief and an escape from work where necessary. Names are not given because of space restrictions, but M. Brown and J. Courage deserve a special mention in this respect. Peter Hayes is thanked for getting all the tables in the thesis typed, especially since this was carried out at very short notice. However, the greatest personal acknowledgements must go to my family and Christine, all of whom have been there when needed and have provided much support and motivation. Christine also patiently proof-read the thesis and therefore deserves a special mention for this, and for giving encouragement when it was most needed. Any acknowledgement of the support given especially from my dad, Jim, and Christine is insufficient to the debt I owe for their personal help.

Finally I thank my mother for the help, love, and support she gave me that goes far beyond this thesis, and I regret I was never able to tell her how much I appreciated it. This thesis is dedicated to the memory of my mother.

Chapter One

INTRODUCTION

1.1 THE INTERNATIONAL CONSTRUCTION ENVIRONMENT

Although the origins of the international construction industry can be traced back to the post World War II era of reconstruction and development of the Colonial states, two factors have emerged within the past 10-15 years that have had a significant effect upon the situation as it exists in the industry in the 1980s. It is argued here that both of these factors are a consequence of the great surge in demand for infrastructure development programmes that were a dominant feature of the 1970s, and can be attributed to increased revenue for construction projects. This in turn was a consequence of the greater prosperity which followed the early 1970s oil price boom and the relatively unrestricted capital market for development oriented loans in the period. The major areas in which the effects of this were felt were Latin America, Africa, the Middle East, and to a lesser extent South East Asia.

The first factor to have emerged in international contracting over the past 10 years is the influx of contractors into the industry from Less Developed Countries (LDCs). (1) This is a direct consequence of the availability of finance for major development projects in the late 1970s, particularly in the Middle East where extensive financial reserves from oil revenues pushed demand for construction projects to a level unprecedented in any region. Backed by home government support, the ability of LDC contractors to undercut Developed Country (DC) (1) contractors on projects requiring little technical expertise but a high input

Introduction

of semi skilled and unskilled labour is almost entirely due to LDC contractors 'exporting' entire construction teams (including unskilled labourers) from their home countries, and so taking advantage of lower wage rates. This has invariably been reflected in extremely low bid prices by LDC contractors that could not feasibly be matched by DC contractors without making a loss. Of this group of LDC contractors the South Koreans are the best known and most successful, (2) but the competition that DC contractors have faced in the Middle East especially from Indian, Pakistani, Eastern Bloc and more recently Chinese contractors has been a significant factor in the contemporary situation.

The second factor to have evolved in the 1980s has been the lower level of demand worldwide for international contractors services; in 1985 the American periodical Engineering News-Record (ENR) reported a 14% drop in foreign earnings of the top 250 international contractors in all developing markets during 1984. Although this can partially be put down to the lower workload that exists in the Middle East as a consequence of the completion of much of the infrastructure, the lack of development finance available for Africa and Latin America as a result of the worldwide recession and dismal conditions in those markets has also had a limiting effect.

When taken together, these two factors adequately summarise the international construction environment in 1985; the fall in demand and increasing numbers of contractors in the industry has led to a state of overcapacity. This in turn has caused an increase in the ferocity of competition and a falling off of profit margins to a level never experienced in the industry. Consequently the situation now involves a number of complex factors, and competition can no longer be characterised by price alone; home and host country political links, sophisticated marketing techniques and incentives, and particularly the ability of the contractor to provide project finance have all become factors that determine success in the industry. The ability to provide a specific project within the time allocated for construction has become overshadowed in the bidding process by the competency of the firm to differentiate its tender from others on the tender list. While the importance of a low price should not be underestimated, other significant factors may contribute, or even exceed this factor, to guarantee success. Clearly factors attributable to the firm are relevant here, but increasingly

Introduction

the size and scale of projects, and the level of finance necessary has meant that ultimately provisions made by home country governments to help contractors compete overseas have become a critical factor in the industry. In some cases this has reached the point where contractors of certain nationalities can attribute their success abroad almost solely to the help provided by their home country government in the procurement of overseas construction contracts. Within this environment contractors from countries that do not provide against extensive government support have faced increasing problems, especially against the quasi nationalised construction industries from some LDCs and DCs (eg France), that compete as an integrated body incorporating finance, access to home country materials and equipment related to the project, and powerful government support as competitive advantages.

In the face of such a competitive and aggressive environment the UK construction industry as a whole has appeared to retain a significant share of the market over the last 10 years. Figure 1.1 illustrates the contribution made by UK architects and surveyors, construction contractors, and engineering consultants to UK invisible earnings as the 'TOTAL' figure. From 1978-83 total invisible earnings from overseas construction activities rose from £646 million in 1978 to £796 million in 1983, an increase of approximately 23%. These statistics however, overshadow the trend that emerges if this total is broken up into the respective sections; Figure 1.1 also illustrates that while UK construction contractors have maintained a steady return, much of the increase in invisible earnings can be attributed to overseas activities of consulting engineers and to a lesser extent architects and surveyors. In 1978 consulting engineers contributed 57% of total UK invisible earnings from overseas construction work, with construction contractors contributing 33%. By 1983 the share of construction contractors had dropped to 16%, while that of consulting engineers had risen to 70%. If real changes of earnings overseas are introduced by comparing percentage change of earnings with UK rate of change of inflation (given by the consumer price index), the situation reflects the contemporary trend of the UK construction industry abroad shown in Figure 1.1. Figure 1.2 illustrates this situation, using the consumer price index as the UK rate of inflation; the 'Total' figure shows that post 1980 the UK construction industry has increased earnings above the rate

Introduction

Figure 1.1: Invisible Earnings of UK Construction Industry Overseas, 1978-83

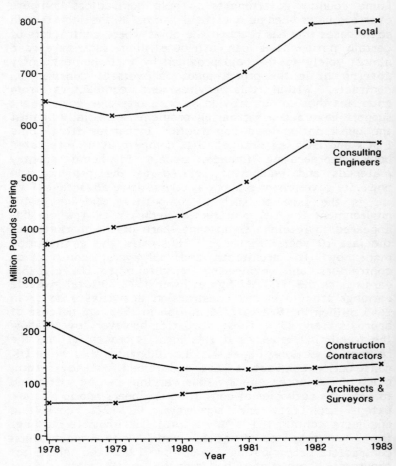

Source: CSO 'United Kingdom Balance of Payments, 1984 Ed.'

of inflation, indicating that real earnings have also increased. Much of this can be attributed to the increases in awards of UK architects and surveyors abroad as well as British consulting engineers overseas earnings, although in 1983 UK consulting engineers experienced a major fall in awards abroad. Figure 1.2 also shows that UK contractors

Introduction

Figure 1.2: Percentage Change in UK Construction Industry Overseas Earnings and Consumer Price Index 1979-83

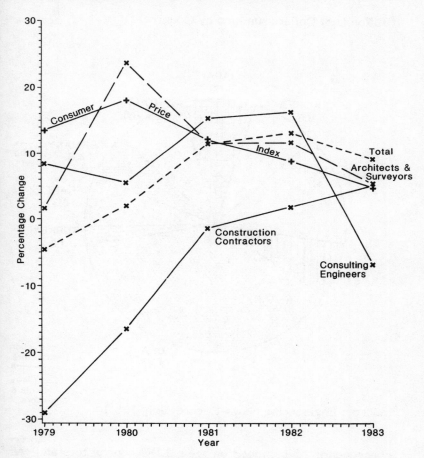

Source: As Figure 1.1 plus IMF 'Financial Statistics' (1985)

over the period have faced a decline in real earnings though the incidence of this has fallen over 1979-82. For the first time in the period analysed, in 1983 UK contractors experienced a real increase in level of overseas awards they received.

The statistics in Figure 1.1 and Figure 1.2 clearly illustrate that although UK contractors have become more competitive throughout the period, they have suffered as a

Introduction

Figure 1.3: Overseas Earnings of Top 250 International Contractors, 1984

(Billion U.S. Dollars summed by Country)

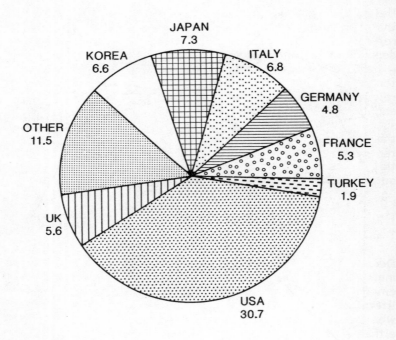

(Source: Engineering News-Record, July 1985).

consequence of competition from LDC contractors and the fall in worldwide demand. However, it is worth emphasising that extensive competition has also come from other DC contractors; in Africa, for example French contractors have pursued an aggressive policy of entering markets traditionally serviced by UK contractors - in 1982 some 33% of French African contracts (approximately £1.5 billion) came from English speaking countries. (4) In general, however, competition for all construction projects in all developing regions has significantly increased as a result of the fall in worldwide demand for major construction works.

Introduction

Within this environment, some DC contractors have done better than others, and these trends appear to be related to the nationality of the construction firms; Figure 1.3 gives a clear indication of this. The chart shows foreign earnings by nationality of the 250 largest international contractors worldwide for 1984, and demonstrates that American, Italian and Japanese contractors have been more successful than UK, German and French contractors in most developing regions. Explanations for this may take several forms depending upon the nationality of the contractor in question - home government support, political and historical links, technical ability, and other related factors may all be relevant. That these trends are national suggests that home and host country specific factors are instrumental in the success of a country's contractors abroad. In view of this, the argument that factors other than price are significant in the winning of bids seems a justifiable hypothesis. Bearing this in mind, it is within the contemporary environment outlined above that this study is based.

1.2 DEVELOPMENTS IN THE ANALYSIS OF THE INDUSTRY

Despite the problems faced by UK contractors in overseas markets, and the potential for detailed theoretical and empirical analysis of the international contracting industry, literature on the subject is not extensive. Furthermore, what studies have taken place have tended to be concerned either with the application of the industry to the nature of development, or the operations of contractors abroad within a management framework as a guide to the way contractors should operate.

Studies of the contribution of international construction to development include the works of Cockburn (1970) and Edmonds (1972). The approach of these and similar studies has been to suggest that the onus of development is placed upon international contractors and developed country governments. While this study supports this general argument, it is not within the limits of this research to incorporate this type of approach to any significant level.

The investigation of factors specific to international construction and the management of international operations, as a second stream of literature on international contracting, has been the subject of numerous articles and

Introduction

features within the trade journals. Useful contributions to this approach include Baker and Cockfield (1984), Cox (1982), Hamman (1971), and Neo (1975). (5) Because these works provide a valuable insight into specific factors they will be referred to later in the text.

It is not a criticism of these approaches to suggest that they do not deal with the 'how, why and where' of international contracting; it was neither the direction they chose nor the objective of these studies to evaluate factors that have been significant in the evolution of the industry. However, it is suggested here that the subject matter is of interest in both a theoretical and non theoretical light, and therefore deserves more attention than it has been given in the past. While it can be argued that developments in economic theory, particularly that of the multinational enterprise (MNE) involvement, go some way to explaining these factors, the lack of empirical work specifically related to the industry suggests that more work is needed in the field. For example the work of Porter (1980) refers to factors within the industry that affect the competitive environment but the reference is neither treated in depth or subject to empirical testing.

Over the past 2-3 years, however, the problems faced by international contractors and issues relevant to the industry have been given more coverage; annual surveys into the industry by the Financial Times have been useful in this respect, along with several articles in the trade journals that have pointed to the disadvantages faced by UK contractors with respect to foreign competition. The most useful empirical analysis of the international construction industry (that forms the basis of the Financial Times surveys) is the annual survey of the industry by the American periodical Engineering News-Record. By listing the top 250 international contractors worldwide by level of foreign earnings, the publication provides extensive background for any investigation of trends in the international construction industry. This thesis draws upon this background as a basis for empirical work carried out within the confines of the research topic. Other more qualitative surveys include a private study by the Hawker Siddeley Group into the reasons behind the falling share of work in developing countries (1985), while publications by the Overseas Project Board of the Department of Trade and Industry have consistently pointed to the problems faced by UK contractors overseas.

Introduction

One area into which fairly detailed analytical work has been undertaken is that of the European process plant contracting industry (Barna et al. 1981, Barna 1983). By investigating factors that have led to the development of the industry, and discerning factors that affect the contemporary competitive situation (including home government support and the ability of firms to adapt to a changing market), the work touches on the research subject of this thesis. Although the work on process plant contracting is limited in scope in relation to this research, and does not rely upon a rigid theoretical model for analysis, the work does provide a useful insight into the industry (which will inevitably overlap with international contracting), and therefore will be referred to later in the text as an integral part of the subject of this thesis.

The work into the process plant contracting industry provides a sharp contrast with that into international contracting; one factor to emerge from this research is that there is a lack of theoretical analysis which tries to justify the contemporary situation in the international construction industry. The 'how, where and why' approach mentioned has not been a feature of past literature and it is argued that this type of analysis would give a better understanding of the trends and problems within the industry today. It is hoped that this research makes a contribution to this end.

1.3 DEFINITION OF TERM

For the sake of clarity it is perhaps useful to define what is meant by the term 'international contractor'; a contractor is herein defined as an enterprise that utilises its productive facilities in predemanded constructional activities using capital that is not owned by the firm. This definition therefore excludes the property development and large scale house building activities that several of the larger European contractors are involved with in other areas of Europe and North America. Firms with process plant interests are included in the definition providing that those firms have facilities for, and undertake, construction activity (whether in relation to process plant projects or not), rather than subletting the work to a construction contractor.

Within the context of this definition, an international contractor is a contracting company that works outside the country in which that company is registered (referred to as

Introduction

the 'home country'). It is suggested here that the major markets in which international contractors will work will be the developing region markets, rather than the industrialised countries of Europe and North America for two reasons:

1. The lack of indigenous contractors in these regions means that clients are in need of the skills and resources that international contractors offer.
2. The potential and actual size of projects in these regions is much greater than in developed countries due to the lack of adequate infrastructure that prevails in developing regions, and therefore these larger projects can support the high overheads and give a return consistent with the higher risk factor which is characteristic of international construction.

1.4 THE RESEARCH FRAMEWORK

The theoretical argument on which this work stands is that economic theory satisfactorily highlights the major issues of the international construction industry in its contemporary setting. It is stated here that the theory of the MNE is the most applicable framework given the objectives of the research for four reasons:

1. The primary factor that motivates construction firms to consider overseas production is the maintenance or improvement of monetary gains (either as a direct or indirect policy measure) associated with such ventures. The framework of the analysis must therefore encompass economic aspects and theory.

2. The international contractor is involved with international production since by definition an international contractor works outside the boundaries of his home country. This leads to specific characteristics and problems linked to international involvement which form the major components of international production theory.

3. The literature of international production theory covers a wide range of other fields of economics under the auspices of the existence of MNEs, including industrial organisation theory, international economics, and location theory. The framework therefore provides a flexible basis from which the industry can be analysed.

4. The origins of the theory lie in the analysis of multinational manufacturing industry, and therefore the

Introduction

application of the theory to international construction (which is essentially a service industry) may yield interesting results both to scholars involved with multinational production theory and those engaged in the industry.

1.5 OBJECTIVES AND HYPOTHESES OF THE RESEARCH

The overall objective of this research is to analyse characteristics of the international contracting industry that have had a major effect upon the firms operating as international contractors and also may have been significant in explaining the contemporary situation in the industry. By attempting this the study also aims to illustrate that theoretical developments in the field of international production and the MNE go some way to providing a framework that has powerful predictive and analytical properties which are useful in explaining trends and features in the industry.

The specific objectives of the research are given as:

1. To examine international contracting within the framework of the received theory of the MNE to see how successfully the theory can be applied to the industry, and to analyse what contributions the theory can make to our understanding of the contemporary situation in international contracting.

2. To illustrate how UK firms operate overseas in terms of motives for international operations, choice of market, and methods of procurement of work and construction activity overseas.

3. To contrast UK contractors' operations overseas with those of other countries contractors to illustrate how they differ in terms of motive, locational choice, and method of international operations, and to investigate whether any differences that occur may be explained by theoretical reasoning.

Within the confines of these objectives and the chosen conceptual framework, the approach aims to:

1. Review the economic literature of international production with the intention of finding a framework that most satisfactorily summarises factors likely to be relevant to international contracting.

2. Outline the characteristics of international construction that differentiate the industry from all others

Introduction

(including domestic construction).
3. Synthesise the industry characteristics with the theory of the MNE as a means of explaining the existence, location, and motives of international contractors.
4. Re-evaluate the major factors relevant to international competition within the industry by means of theoretical discussion, prediction, and subsequent empirical testing, and illustrate where possible those factors significant to the industry which came to light as a result of this process.
5. Yield conclusions based upon the research that may be justified using the adopted approach.

On the basis of this framework there are three hypotheses that form the foundation of the research as a theoretical and empirical investigation.

> **Hypothesis 1:** The construction industry may be differentiated from other industries by specific characteristics of the industry that occur due to the nature of production which itself is a factor of the final product of construction activity (eg bridge, dam, etc.). In addition to this, international construction activity can be classed as a separate industry to that of domestic construction because the fact that international contractors work in an alien environment causes a number of additional problems unique to multinational involvement of an enterprise.

The implication of this hypothesis is that if the international contracting industry does exhibit patterns of behaviour that are attributable to industry specific characteristics, then results of the study may yield conclusions that are different to other industries. If this is not the case, there is no valid argument for treating international construction as a special case.

> **Hypothesis 2:** The theoretical framework of the nature of the MNE as outlined in this research can be legitimately applied to the international contracting industry in spite of specific characteristics of the industry and despite the fact that the theory was initially formulated to explain multinational manufacturing.

Introduction

The additional complications faced by an enterprise which undertakes multinational production form the foundations of international production theory, and the implication of this second hypothesis is that contractors face similar problems to companies from other industries because of overseas involvement. However, when taken in conjunction with the first hypothesis this leads to the possible conclusion that industry specific characteristics will influence the contractors situation. Therefore if both hypotheses are valid we would expect international contractors to modify behaviour according to determining factors that relate to the industry. Hence it is not necessarily the case that the actions of international contractors will be similar to multinational manufacturing firms for example, despite the existence of a common core of problems associated with international involvement.

Hypothesis 3: The application of the theoretical framework to international contracting provides explanations concerning the industry that are of academic and practical importance. In particular the theory is structured in such a way that it will set out justification and explanation for differentials that occur in the industry at firm and country specific (6) levels with regard to method and motive of operation, competition and locational policies of the firms.

This third hypothesis may be regarded as a logical succession of hypotheses 1 and 2. Stated simply, if the industry has specific characteristics that can be encompassed by the theoretical framework then that framework should provide reasoning for the prevailing situation. This reasoning is partly derived from the nature of the theory which enables this process to be identified.

A consequence of the validity of these hypotheses is that if they are correct then the conceptual framework utilised may be an effective tool for analysing reasons for whatever differentiation occurs in the industry and may be used as a basis for any remedial changes of firm or country specific factors that could improve the contractors competitive standing. The decline in the competitiveness of UK contractors worldwide, for example, may be investigated using such a process. This may be added as an underlying objective of this study.

Introduction

1.5.1 **Empirical Setting**

As stated within the framework, this research is an attempt to see how closely the theory applies to the actual situation, therefore some empirical work is an essential part of the analysis. This empirical investigation was carried out in two ways:

1. Meetings with people connected with the industry. This included contributions from several institutional bodies, consulting engineering firms, trade magazine journalists, international and merchant bankers, and civil service personnel.
2. Interviews with executives from contracting companies presently engaged overseas as international contractors.

Although both aspects provided a greater understanding of the contemporary environment of the industry, the empirical results presented in this study come mainly from discussion with executives of major contracting companies. These interviews were structured around a questionnaire designed specifically for the research which focused upon issues that emerged from this analysis. In addition to this, company reports (where available) provided quantitative information relating to the firms interviewed. The results of this empirical work are presented later in the thesis.

1.6 FOOTNOTES

1. See World Bank, 'World Development Report 1984' (1984) for a synopsis of DC and LDC (referred to as 'developing') countries.
2. In 1984 Korean contractors within the top 250 international contractors worldwide took fourth place in terms of amount of foreign earnings by nationality group, behind the Americans, Japanese and Italians. (Source: Engineering News-Record, July 1985).
3. Financial Times, 5th March 1984.
4. Neo has perhaps come closest to attempting to illustrate factors within the industry that encompass the objectives of this study, but a criticism that can be levelled at Neo's work is that he assumes that the provision of finance and other potential competitive advantages are not

Introduction

affected by the nationality of the contractor. This is a serious fault within the confines of the present research which argues that country specific factors may be extremely relevant. This point will be returned to later in the text.

5. By country specific here we mean analysis of contractors according to their nationality in order to investigate whether there are national trends affecting the industry. However, the analysis may include (where stated) investigation by market or region to discover trends that emerge therein and also investigation of factors important to the individual firm (ie firm specific factors).

Chapter Two

A LITERATURE REVIEW OF MNE THEORY

The focus of the economic theory that provides the framework for the analysis presented in this research is that of developments in the theory of international production and the Multinational Enterprise (MNE). An MNE is herein defined as an enterprise that undertakes Foreign Direct Investment (FDI) in order to engage in production in countries other than the country in which the firm is registered (i.e. the home country). Following Dunning (1974) any firm that engages in FDI is an MNE in the context of this work.

The analysis of the MNE as a separate field of economic theory has evolved as a consequence of the growing importance of MNEs in the world economy; the post World War II surge of FDI and the growth of MNEs has produced such an increase in the involvement of the MNE in world trade over the period that Dunning (1974) has suggested that 'MNEs are among the most powerful economic institutions yet produced by the private enterprise system.' (p. 13, 1974)

Empirical studies tend to substantiate this - in 1978 a UN commission concluded that MNEs accounted for one fifth of the world's output in the mid 1970s (excluding Eastern Bloc countries), while Dunning and Pearce (1981) found that approximately 27% of the sales of the 866 largest industrial enterprises worldwide were earned through foreign affiliates. Within the context of this, economic theory has developed to explain the emergence of the MNE given that there are characteristics peculiar to firms working in more than one national environment. The basis of the developments revolve around the questions of 'why,

A Literature Review of MNE Theory

where, and how?' as applied to MNEs.

Until the 1950s, much of the economic theory was based upon the Classical theory of international trade. The application of this to the development of the MNE was constrained by the limiting assumptions of the framework; the Heckscher-Ohlin model of international involvement, for example, argued that international trade based upon comparative advantage would ensue if labour and capital were homogeneous inputs of production that were geographically immobile; all enterprises were price takers; barriers to trade and transactions costs were zero; production functions were efficient and internationally identical; and international tastes were alike. Clearly assumptions with this level of rigidity prevented the application of trade theory to the MNE.

The inability of international trade theory to explain MNE involvement, plus the fact that an MNE may be classed as an exporter of equity capital as a consequence of setting up foreign subsidiaries, was the motivation behind economists in the pre-1960s considering FDI exclusively as a form of international movement of capital. Within the framework of international capital movement theory, it was argued that differential interest rates of capital between countries led to the flow of portfolio and direct investment from capital abundant to capital poor countries; this argument rests on the theoretical basis that different interest rates of capital between countries are a product of different marginal efficiencies of capital that reflected the relative endowments of capital internationally. Hence a capital rich country will have a low efficiency rate while a capital poor country will have a high rate, leading to a flow of capital to the poor country to exploit higher interest rates on capital.

The similar determinants and characteristics of portfolio and direct investment, and the parallel conditions under which they supposedly occur, led many economists to critically examine the basis of this 'capital arbitrage' hypothesis. As a consequence of this Hymer (1976) set out the distinction between portfolio and direct investment which remains the crucial factor separating the two:

> If the investor directly controls the foreign enterprise, his investment is called a direct investment. If he does not control it, his investment is a portfolio investment. (1960, published 1976).

This idea, that FDI involves the control of the invested capital, forms the basis of the definition of direct investment that is used in this research. Following on from this definition it can be argued that FDI will not be an efficient way to transfer capital between countries because of the additional costs that characterise overseas production. Hymer has argued that the capital arbitrage hypothesis is inconsistent with several features that emerged in the pre-1960s era concerning the behaviour of MNEs.

1. Historically capital flows to and from the United States have shown net imports of portfolio capital and net exports of FDI. Clearly therefore the motives for portfolio and direct investment cannot be put down to a common single determinant that in this case is supposedly interest rate differentials.

2. Some countries have been both FDI investors and recipients. If MNEs are involved in capital arbitrage, this would suggest that in these countries interest differentials occur between industries. Hymer argues that while this may partially be true, the movement of interest rates through time have not changed the patterns of industries persistently associated with MNE involvement. Therefore this rules out the hypothesis that industry specific direct investment is a consequence of inter industry interest rate differentials.

3. If the capital arbitrage hypothesis were correct it would imply that the majority of enterprises undertaking FDI should be large financial firms. However, the dominant feature of FDI in this respect is that it is characterised by what Hymer referred to as 'nonfinancial firms'. This obviously goes against the capital arbitrage hypothesis.

These factors taken together suggest that the capital arbitrage hypothesis is not valid as an explanation of the movements of or motives for FDI. This result has been significant in the development by economists of other hypotheses concerning the existence of the MNE that have originated in other fields of economic theory such as industrial organisation theory and location theory. It is to major developments in these and related fields that this survey now turns.

2.1 BEHAVIOURAL ASPECTS OF FOREIGN INVESTMENT

Of the behavioural models that apply to the MNE, the work of Aharoni (1966) is possibly the most extensive. According to Aharoni, the initial decision to invest is stimulated by chance motivation within the enterprise in response to internal management interest and external factors such as client demands, tariff wall increases or oligopolistic reaction. Time and cost spent on appraisal will be positively linked to probability of the firm undertaking FDI as individuals within the corporation become more interested and motivated by the initial idea. Aharoni argues that this reasoning, and evidence from his case studies rejects any profit maximisation motives for undertaking FDI, because the firm does not systematically use economics in a logical framework of appraisal and subsequent direct investment.

Criticisms of Aharoni's conclusion can be summed in three points:

1. As Stevens (1974) has indicated, Aharoni's conclusion is not valid since his approach may be compared to a firm operating profit maximising conditions under uncertainty; if the risk adjusted costs of search are greater than expected returns the company would not initially undertake appraisal. Hence an economic basis to the appraisal of opportunities is evident.

2. Buckley and Casson (1976) have suggested that Aharoni's case study does not give a good representation of companies since the firms interviewed were new to the international scene and therefore unable to utilise a procedure for investment appraisal based upon experience. This falls in with Dunning's criticism (1974) that Aharoni's behavioural procedure is inefficient though basically organised. It is likely that these inefficiencies would be greatly reduced with the experience of further ventures that would possibly lead to the emergence of profit maximisation factors as a primary motivation for foreign investment.

3. Aharoni ignores the initial stimulus to going overseas, which is essentially a product of economic factors. Stopford and Wells (1972) have argued that managers may be motivated to go abroad for prestigious reasons, but that potential profitability abroad remains the major concern. Aharoni largely ignores this prerequisite, and therefore ignores the fact that overseas involvement may be the only way a firm can fully develop its internal skills (Penrose, 1956). Aharoni thus fails to explain why some firms may be

successful abroad while others may not, which it is argued here is a crucial factor in the study of the MNE. Because of this, the work of Aharoni has only limited relevance to the study of the MNE within an economic framework.

2.2 INDUSTRIAL ORGANISATION THEORY

Much of the work into MNEs in the 1950s and early 1960s centred on the question 'how is it possible for MNEs to compete with indigenous producers given the additional costs of doing business abroad?'. This led economists to adopt an industrial organisation approach to suggest that MNEs have characteristics that give them a net comparative advantage over other firms supplying the same foreign markets. The seminal work in this field is considered to be that of Hymer (1960, published 1976) who essentially argues that Bain's (1956) notion of barriers to entry in domestic oligopolistic markets could be equally applied in an international context; the ability of the MNE to compete in foreign markets against established indigenous companies could be attributed to the MNE being able to acquire and sustain competitive advantages over those indigenous firms. These advantages would enable the MNE to overcome what Hymer considers to be the two major disadvantages of entering an alien market:
    1. Local firms have knowledge of consumer tastes, local business customs, legal framework, and will be treated favourably in a situation of xenophobic or nationalistic prejudices;
    2. The foreign firm incurs additional costs of operating over those faced by the local firm as a consequence of operating at a distance from the parent or controlling company. These costs include travel, communication, and time lost in addition to costs of errors that occur because of the lower quality of internal communication in the firm that may be positively related to distance.
    By utilising the competitive advantages available to the MNE Hymer argues that the enterprise cannot only vault these barriers to entering a foreign market, but also erect barriers against indigenous producers and other MNEs to form an oligopolistic or monopolistic market in that country.
    While ownership of advantages provides a necessary

condition for FDI in Hymer's analysis, it is nevertheless not a sufficient one to explain the existence of MNEs; the ownership of advantages does not suggest why the firm should undertake FDI rather than export its product or licence its advantages in foreign markets, since in theory these would provide the same economic rent without incurring the additional costs of overseas involvement. Hymer has suggested that FDI will take place in these circumstances where the return for internal utilisation of advantages abroad by the firm are greater than external usage. The implication of this second condition of Hymer's work is that FDI will take place only where the external market for sale of competitive advantages is not efficient, which it is suggested by Hymer will occur particularly in markets for patents and other forms of knowledge. Hymer's ideas have been adapted by Kindleberger (1969) to suggest that FDI would not exist in a world of pure competition. Kindleberger has argued that if all markets operated efficiently, external economies of production and marketing were non-existent, information was costless and there were no barriers to trade or competition, international trade would be the only possible form of international involvement. From this standpoint, the nature of monopolistic advantages that produce FDI can be indicated under four basic headings. First, departures from perfect competition in goods markets, including product differentiation, special marketing skills and administered pricing; second, departures from perfect competition in factor markets, including access to patented or proprietary knowledge, discrimination in access to capital and human skills embodied in the firm (mainly management); third, internal and external economies of scale, including vertical integration advantages; and fourthly government intervention particularly on restriction of output or market entry.

The fundamental basis of this so-called Hymer-Kindleberger (H-K) theory, i.e. that the firm owns and utilises monopolistic advantages to compete in foreign markets, has led to subsequent analysis of the nature of the advantages on which FDI may be based. Caves (1971, 1974) has pointed to product differentiation; Aliber (1970, 1971) has emphasised advantages related to nationality held by firms in particular currency areas; Johnson (1970) has stressed that the public good nature of knowledge confers an internal cost advantage to the MNE; Hirsch (1976) has suggested that Research and Development (R&D) provides a significant advantage while Mason (1975), following Kindleberger suggests that access to factors on favourable terms is the most important advantage.

An implicit factor that is a necessity for any theoretical industrial organisation approach to FDI is the mobility of competitive advantages. This aspect remained largely ignored until Lall (1980) introduced the notion that immobility may be the determining factor in the firms' choice of expansion into foreign markets. Lall has suggested that Hymer's competitive advantages may be firm-, industry-, and/or country-specific, and that the extent of foreign involvement depends (given lower production costs abroad) on the transferability of those advantages from the home country; MNEs will predominate in industries that have highly mobile competitive advantages.

The work of Lall has two important implications. Firstly, it suggests that not all industries in all countries will undertake FDI, but that physical characteristics of competitive advantages will determine the extent of foreign involvement. Secondly, monopolistic advantages will be different according to firm-, industry- and country-specific determinants. While this analysis has made a significant contribution to the validity of the industrial organisation approach, there are still several points of criticism concerning the application of industrial organisation theory to MNE activity:

1. By ignoring locational factors the industrial organisation approach fails to predict not only where advantages will be exploited, but also the initial motivation of the firm to undertake overseas operations and the interaction between home and host country factors that may affect the firm's advantage. As such, the approach cannot be said to sum up all the relevant factors in the decision to undertake FDI.

2. An established MNE may not face high 'costs of foreignness' due to experience at adapting to local situations and international demand for the firm's product. A further point is that DC firms may not experience any effective competition in LDC markets, and therefore entry barriers will be minimal. The importance of competitive advantages may therefore be overexaggerated.

3. The H-K framework does not give a valid explanation as to why firms invest in competitive advantages rather than some other type of asset.

Advantages are assumed to exist in the theory and therefore predictions about initial growth of the firm, and any subsequent investment in advantages once the firm has gone abroad are neglected. This makes it difficult to use the theory to explain maintenance of share of overseas markets.

4. Buckley (1983) has argued that the notion of firm specific advantages relies upon a specific set of assumptions about:
   a. the diffusion of technical and marketing know-how;
   b. comparative advantage of firms in particular locations;
   c. existence of particular types of economies of scale.

Dynamically these factors are likely to change and lead to a decline in the effectiveness of competitive advantages that is not catered for within the static framework of the H-K analysis.

2.2.1 Aliber's Theory of Currency Differentials

Aliber's theory of the MNE is in line with the H-K tradition in the sense that he explains FDI in the context of the MNE possessing an advantage over host country competitors. However, Aliber's approach indicates that this advantage is not firm specific, but applies to all firms in a particular currency area.

The central issue in Aliber's analysis is the currency premium. This premium refers to the rate of interest of debt denominated in any currency, which will be higher the less stable the currency as an incentive to investors in that currency. Aliber argues that investors are myopic, ie that they treat all assets of the MNE as if they were in the same currency as the parent firm, so that for example a US company's factory in the UK will be regarded as a Dollar asset.

Aliber's argument is that source country firms will come from highly stable, low currency premium countries, who can realise an immediate profit from takeover of a less stable currency denominated asset (at a higher indigenous rate of interest) either in that host country or in a third location, and therefore the higher capitalisation rate of the source country enterprise provides the impetus, and means of MNE involvement.

The implications of Aliber's model are threefold:

1. FDI will originate in countries with high capitalisation rates and flow to countries with relatively low capitalisation rates.
2. FDI will occur in those industries where capital expenditure plays an important part in the development of the firm; hence MNEs will be prevalent in R&D-intensive and capital intensive industries.
3. Within Aliber's model, because advantages are non-monopolistic, the foreign firm has an advantage over indigenous producers only, and not over other firms of the same nationality. A corollary of this is that where there are no indigenous producers, Aliber's model suggests that foreign firms need not have any kind of advantage relative to the other firms actually operating in the industry. As such, Aliber claims that this approach replaces that of the H-K framework.

While Aliber's theory may be strengthened by the fact that it predicts the post-war expansion of MNEs - particularly US takeovers in Europe in the 1950s and 1960s, and the success of Swiss and Dutch MNEs worldwide, several criticisms of the approach should be outlined.

Firstly, while it is possible that investment opportunities may be increased by interest rate differentials, to claim as Aliber does that this theory replaces the H-K tradition is overstating the case. The theory put forward by Aliber is more a variant of the H-K theory than an alternative since there is no reason why Aliber's conditions are sufficient or necessary to invoke FDI; the situation described by Aliber could be attributed to technical know-how and the large size of the domestic market in the source country, as easily as to currency differentials, with the same results.

Secondly, Aliber fails to explain the cross-hauling of FDI that is a phenomenon of the 1970s and 80s. Within Aliber's framework cross-hauling could be explained by subjective capitalisation rates, but this suggests that the whole motivation for FDI in Aliber's theory is based upon risk, and introduces the possibility that currencies may be prone to distortions which reduce their efficiency as price signals. In a risk averse situation (which reflects Aliber's argument) this would limit finance for overseas projects significantly.

A third point, in connection with the second, is that Aliber's theory is circular and implies that ultimately there will be only one source country of FDI, with all others recipients; MNEs from the source country should gain from a strong currency, and the currency will get stronger as a result of MNE success in overseas markets. The existence of cross-hauling of FDI suggests that this is not the case, and it is argued here that Aliber fails to incorporate other effects of currency movements on the domestic environment which will have a significant effect on the actions of that country's firms abroad.

Fourthly, Buckley and Casson (1976) have argued that the condition that portfolio investors are myopic, which forms a major part of Aliber's analysis, may be in doubt if one considers that institutional investment managers are unlikely to think along these lines in the international capital markets of the 1970s and 80s. Furthermore, Aliber fails to explain 'greenfield' FDI, or capital flows within currency areas (eg the investment of US firms within the Dollar area). In the wake of these criticisms, the validity of the Aliber model as an explanation of the motivation behind movements of direct investment should be treated with caution.

2.2.2 Caves' product differentiation theory

Whereas Aliber stresses non-monopolistic competitive advantages within the H-K framework, the work of Caves (1971, 1974, 1982) is more in line with the initial work of Hymer in emphasising traits of monopolistic competition. Caves has suggested that the main advantage that MNEs have over their local competitors in foreign markets is their ability to differentiate products. In the respect that this requires knowledge from the MNEs' domestic experience that is used at little additional cost overseas, Caves adopts an approach similar to Johnson (1970), even though Caves clearly emphasises product differentiation rather than Johnson's analysis of technology exploitation as the major competitive factor.

Caves has argued that the MNE exploits product differentiation either by differentiating the same product across different regions, or differentiating a wide range of products to the tastes of one region, or possibly a combination of the two. The former strategy leads to horizontal diversification while the latter leads to conglomerate diversification. However, within the analysis Caves does not suggest that product differentiation will always mean the servicing of foreign markets by FDI. Caves classifies three possible groups of differentiated products: If the cost of the product benefits from economies of scale of production, and the product requires little adaptation to local market conditions, then exporting of the product will ensue; if the product does not enjoy economies of scale, or if the product involves a proprietary process, licensing of foreign firms may occur; however if the firm's major competitive advantage is embodied in the marketing, research and management skills rather than the existing product then FDI will take place.

A criticism of Caves' approach is that his model does not cater for other types of competitive advantage by relying upon product differentiation, and therefore is limited in its application. Possibly a more important argument is that Caves' analysis of how markets may be serviced is simplistic and fails to recognise the implications of costs that may be associated with externalisation of the firm's competitive advantages. The theory of the firm suggests that the decision to externalise the firm's assets (eg by licensing) depends upon the relative costs of different contractual arrangements while the theory of trade suggests that this must involve no cost at all. By utilising these approaches together Caves fails to recognise that they could only be compatible if the best contractual arrangement is costless, and therefore does not give a valid explanation of the motivation and ability of the firm to externalise its advantages in foreign markets.

2.2.3 The Product Cycle Theory

The Product Cycle (PC) theory put forward by Vernon (1966, 1971, 1974) can be called an extension of the industrial organisation approach to FDI, based upon product differentiation with a time lag. Vernon's contribution remains the innovative treatment of trade and investment as part of a historical process of exploiting foreign markets that introduces a dynamic dimension to MNE theory.

Following criticisms of an earlier (1966) version of the model, Vernon (1971, 1974) revised the concept of the PC theory to include aspects of oligopolistic behaviour though the framework remained for the most part intact. In the mark II PC model there are three distinct stages of the cycle:

A Literature Review of MNE Theory

1. Innovative oligopoly stage. As a consequence of socioeconomic development and oligopolistic competition, firms in advanced countries undertake extensive R&D to produce labour saving producer goods as a response to their comparative advantage in capital that is reflected in relative price of the factors of production. Vernon argues that this will initially take place in the most advanced market (ie where income levels are highest) where communication costs and distance of mutual market signals between producer and consumer are lowest. The theory suggests that location of production in this case will take place near to the market (ie the home country of the producer) to minimise communication costs and distance and so hasten the continual process of adaptation and improvement of the product and production process that will occur through communication between production, marketing and consumer demand for the new product.

2. The Mature oligopoly. Barriers to entry of economies of scale of production, marketing and R&D form effective constraints upon competition in the industry so that oligopolistic defence and expected response to rivals' actions become the major issue. To maintain domestic market share, all firms seek out potential markets worldwide in which they may gain a significant advantage over home country competitors via intercountry cost differentials or advantages accruing from multinationality. This leads to a 'bandwagon' or 'follow the leader' defensive movement by rivals, who locate in the leader's overseas markets to stabilise world market shares of the firms so that rivals strengthen or maintain their bargaining position with respect to competitors. Stability is reached when each of the rival firms produces in each of the world's major markets.

3. The senescent oligopoly. Economies of scale break down as an efficient barrier to entry, and after attempts to erect artificial barriers (eg product differentiation through advertising) some firms leave the industry altogether while others stay on and reconcile themselves to a competitive environment. In this case location cost differentials play a major role in choice of location.

Although the theory provides a useful explanation of the firm's competitive strategy and locational factors, the framework has been criticised on several grounds. A primary objection to the theory is that the dynamic nature of the product cycle does not cater for rate of change of innovation or time lags; while the PC theory predicts the sequence of events, it does not suggest how soon the next stage of development may occur, or explain exogenous effects of other firms' products on the standardisation of the product. The PC is thus a considerable simplification of the dynamic process. This point is further emphasised by a second criticism that the PC fails to account for possible interaction between the mature product and developments made by the firm with respect to innovation of the old product to maintain competitive position. If repeated successful innovations can be made, it is argued here that the standardised product may never ensue, but that there be continual movement between the new product and maturing product stages.

2.2.4 **Oligopolistic strategy and the MNE**

Following on from Vernon's PC revisions, Knickerbocker (1973) suggested that in oligopolistic industries, there may be strategic responses to home country rivals' overseas investment. The basis for Knickerbocker's work is that FDI may confer an advantage upon the initial firm that enters a market via FDI rather than export, and that rivals in an oligopolistic industry will follow that leader into these markets by also undertaking direct investment to effectively spoil that market for the leader and maintain competitive market share. This argument suggests that the followers of the leader enter the market to prevent any advantage in the home market accruing from FDI and therefore the strategy is defensive. The thesis that Knickerbocker tested upon this argument was that initial investments in foreign enterprises in a given market and industry tend to be 'bunched' in time, and that this bunching will be greater the more oligopolistic the industry. Knickerbocker's empirical analysis supporting this argument takes three forms: firstly, Knickerbocker shows a positive correlation of bunching across industries with the four-firm concentration ratio, where the ratio acts as an index of oligopolistic performance; secondly, the empirical analysis illustrates a strong correlation of bunching with profitability; thirdly the empirical work also shows a strong correlation between bunching and an index of stability and cohesion of the national market. From these results Knickerbocker argues that oligopolistic reaction is a significant feature of MNE activity in many industries.

This approach however is not without its criticisms. Possibly the most significant of these is that Knickerbocker's results support an alternative argument that the actions of these firms correspond to profit maximising firms acting in a situation where there are imperfections in market information; the 'follow the leader' principle may be a response to firms following the enterprise with the best market intelligence system in order to exploit any changes in the market that may occur, rather than oligopolistic strategy. This would give the same empirical results as Knickerbocker's work even though the argument here is that firms bunch not in response to oligopolistic strategy but as a factor of market information acquired through monitoring the movements of other firms.

A second criticism of Knickerbocker's analysis that is related to the first is that Knickerbocker fails to include both the domestic and foreign market conditions in the framework. This of course is an integral part of the criticism above, but may be highly relevant in some industries. For example, if demand is falling in the home market but clearly rising in a foreign market with the potential for long term growth, it is likely that bunching will occur in that foreign market as a response to demand. This is a product of locational rather than market intelligence factors (although the two may be closely related) but clearly demonstrates that locational factors cannot be ignored as the motivation for entering markets, even when oligopolistic strategy motivation is assumed.

The choice of markets within an oligopolistic framework becomes even more complex if we include the possibility of cross-hauling of FDI as an oligopolistic response. Graham (1974, 1978) suggests that the entry of a foreign firm into a stable oligopolistic national market may cause retaliatory entry by those host country oligopolistic firms to the foreign entrant's home market to dissuade entry or set the grounds for collusive bargaining. This approach may be criticised along similar lines as Knickerbocker's for failing to take account of locational factors, but possibly a more serious criticism is that Graham fails to explain why firms enter each other's markets if rivalry behaviour is expected, or conversely why firms with rent yielding assets do not exploit them to the full via multinationality except in cases of rivalry. This criticism does not suggest that Graham's analysis is invalid, but rather that his conditions for intra industry FDI (IIFDI) are extremely general to suggest that rivalry behaviour may be the norm in oligopolistic industry.

The work of Norman and Dunning (1983) and Casson and Norman (1983) goes some way to alleviating the problems of generality that occur in literature on MNE oligopolistic reaction by specifying conditions under which FDI will prevail in oligopolistic situations. Norman and Dunning argue that the motivation behind FDI in oligopolistic industries may occur when there are ownership specific advantages accruing to coordination of activities under common ownership, and where goods transacted are near perfect substitutes in production and consumption and therefore depend upon firm specific advantages such as product differentiation, brand name, and marketing expertise. Following on from this line of argument, Casson and Norman suggest that it will be difficult for MNEs from one country to control entry of firms into the industry of other nationalities (a) in large markets, (b) in markets characterised by a sophisticated economic and social environment, and (c) for products produced from a reasonably standardised technology, as these factors will tend to increase the oligopolistic uncertainty and increase the motivation to serve foreign markets by FDI. In terms of the Graham model, this analysis has specific relevance; when the industry environment is similar to that outlined above by Norman and Dunning (which is due to the nature of production and the product) and conditions exist as in the Casson and Norman analysis, intra industry FDI (IIFDI) may occur though it is unlikely that it will be a rivalry response since the market will be wide open to foreign entrants and therefore the domestic oligopoly will be minimal. However, if existing conditions are the opposite of those given by Casson and Norman, rivalry behaviour may possibly occur.

An implicit argument in the Casson and Norman work is that the degree of multinationality of an industry will reduce the potential for the erection of oligopolistic barriers to entry. Along similar lines Casson (1983) has suggested that collusion between multinational oligopolists is least likely in industries 'where products are varied, technology and management is progressive and entry from outside is likely' (1983 pp.89-90).

The suggestion here then is that oligopolistic behaviour may be a relevant factor, as suggested by Knickerbocker and Graham, but that this is not the case in all multinational industries; factors such as nature of ownership advantages, product characteristics, state of the market (both international and domestic) and ability to maintain oligopolistic barriers are the key issues in determining the extent of oligopolistic strategy as the motivation behind FDI. Without specific assumptions concerning these factors oligopolistic analysis into MNE involvement remains generalised beyond the point of usefulness.

2.3 INTERNALISATION THEORY AND THE MNE

While the industrial organisation approach is an attempt to answer the question of how MNEs compete abroad, it does not address the more fundamental question 'why do MNEs exist at all?'. The majority of work in this field has centred upon the notion of 'internalisation'. The essence of the concept is that the actions of firms can replace or augment the market process via internal organisation rather than the external market mechanism. By highlighting the costs and benefits to the firm of taking this action, internalisation has become a powerful tool for explaining the existence of firms in general, and provides the backbone to the application of the framework to explain why MNEs exist.

The work of Coase (1937) is usually taken as the seminal work in this field, although contributions by Kaldor (1934) and more recently Penrose (1956) have been important to the development of the analysis. The work of Coase is often cited because Coase demonstrates that a firm may in certain circumstances substitute coordination of production by internal planning of the firm for coordination by prices in the market. The implication of this to the analysis is that internal coordination 'supersedes the price system', and that if this were not possible then there would be no incentive for the firm to internalise production, and therefore no reason for the firm to exist.

This approach, which seems deceptively simple, underlies the argument for internalisation - by having the ability to organise internal coordination of production cheaper than external price transactions, the implicit argument is that not all markets are efficient. Buckley and Casson (1976) have pointed out that '... under certain conditions ... the coordination of interdependent activities by a complete set of perfectly competitive markets cannot be improved upon' (p.36, 1976) which suggests that internalisation will only occur in a state of imperfect markets.

Following Kaldor's (1934) notion of the entrepreneur as the coordinator of production factors, Coase argues that managerial control originates from the employment contract which substitutes a single large transaction for many separate smaller transactions and thereby economises on the costs of using the price mechanism. The suggestion here is that there are costs to the external market that may be minimised in the internal organisation, especially the avoidance of uncertainty concerning the identification of market prices:

> The most obvious cost of 'organising' production through the price mechanism is that of discovering what the relevant prices are. This cost may be reduced but it will not be eliminated by the emergence of specialists who will sell this information. (pp.390-1, 1937)

Within this analysis, the price mechanism will be superseded until costs of internal organisation equal those of the price mechanism.

Although Coase's approach has been criticised for being almost tautological (Alchian and Demsetz, 1972 p.783), possibly a more valid argument is that Coase fails to recognise the potential for entrepreneurial profit within internalisation. It may be argued that the impetus for the entrepreneur to internalise is to gain internal control over pricing of factors of production, and thereby take part of the economic rent that would be paid to the factor in the external market. This criticism however raises a further impetus to internalisation rather than questions the validity of the approach.

Williamson (1975) has suggested that Coase's analysis is incomplete in that it fails to explain why the internal hierarchy does not fully supersede the external market over time, but Williamson may himself be criticised in this respect for failing to recognise Coase's 'decreasing returns to the management function' (1937, p.397) that may push costs of internal organisation above that of the external market. In general, however, the work of Williamson on 'markets and hierarchies' (1975) has given weight to the Coasian approach. Along similar lines to Coase, Williamson has argued that costs in the market are a function of uncertainty, but introduces the notion that these occur because of opportunistic behaviour within the market. Williamson suggests that there are three fundamentals that raise costs in the market with respect to internal organisation:

1. Bounded rationality which reflects the inability of potential transactors to obtain unlimited knowledge to make decisions concerning transactions, due to sociological and language limitations.

2. Information impactedness in the market, where information necessary to transactions is known to one or more parties but cannot be obtained or displayed to others without cost.

3. Opportunistic behaviour as a consequence of the other factors.

Williamson reflects that deprivation of strategic behaviour (ie opportunism via gaming, cheating or shirking) and the exchange of well defined goods and services may minimise market costs to the point where the price mechanism may ensue. However, where transactions are likely to be extended over time or refer to ambiguously defined goods and services, and opportunistic behaviour is not restricted, enforcement and monitoring costs may become prohibitive. The analysis argues that under these conditions the firm will tend to substitute for the market on the premise that the firm's internal control procedures are better suited to organise transactions.

Williamson's work is similar to Coase's in that both suggest that the market versus internal organisation is a result of market imperfections, and while it is not systematically outlined in Williamson's argument, both suggest internalisation is a departure from the complete set of perfectly competitive markets of Classical economic theory. One thing that Williamson does emphasise however as an extension of the original work of Coase is that dynamic change, and most notably technical change, are the major factors that cause inefficiency in the market system (1971, 1979). As such these form the basis of internalisation in Williamson's market vs. hierarchies analysis.

2.3.1 Internalisation within the MNE

Following on from the theory of internalisation, Buckley and Casson (1976) have argued that the link between internalisation and the existence of the MNE is that 'an MNE is created whenever markets are internalised across national boundaries' (p.45). Buckley and Casson suggest that four main groups of factors are relevant to the internalisation decision:

1. Industry specific factors, relating to the nature of the product and the structure of the external market.
2. Region specific factors, relating to the geographical and social characteristics of the regions linked by market.
3. Nation specific factors, relating to the political and fiscal relations between the nations concerned.
4. Firm specific factors, which reflect the ability of the management to organise an internal market.

While firm specific and industry specific factors play an important role in the theory of internalisation as already discussed, the absorption of region specific and nation specific factors into the analysis introduces spatial considerations as a significant factor. Buckley and Casson suggest that locational factors become an integral part of the firm's decision to undertake multinational operations:

> ... the locational strategy of a vertically integrated firm is determined mainly by the interplay of comparative advantage, barriers to trade, and regional incentives to internalise; the firm will be multinational whenever these factors make it optimal to locate different stages of production in different nations. (p.35, 1976)

The choice between internal and external organisation within the Buckley and Casson framework follows the work of Coase in that there are both benefits and costs to internalisation, and that the relative dimensions of each determine the internalisation decision. Casson (1979) has suggested: 'Profit maximising firms will internalise markets up to the margin where the private benefit is equal to the private cost.' (p.55, 1979).

As an extension to the initial Buckley and Casson analysis, Casson suggests that there are four separate benefits and four separate costs of internalisation that predict where the operations of MNEs will be concentrated in particular areas of economic activity. The first benefit of internalisation may occur in the market clearing system. The internal organisation provides greater flexibility for market clearing than in the negotiation process of external markets; higher costs of communication internally will be offset by the avoidance of reiteration of trial prices in the market that permits a faster response to changes in the economic environment. Furthermore, discriminatory pricing is not necessary under internal organisation and therefore benefits over external markets under increasing returns to scale by not having to take account of competitors, transactors or government bodies that could prevent this form of pricing in the external market.

Secondly, in terms of contractual costs there may also be internalisation benefits; the opportunistic incentive to gain via contractual default or error is a problem of the external market that is non-existent in the internal organisation (Williamson, 1975). Internal organisation may therefore be preferable in future contracts (as a consequence of uncertainty), and complex intermediate products that cannot adequately be protected in a contract from default and exploitation.

Thirdly, adequate property rights may not be clearly assigned. In cases where ownership rights are not well founded in law, the internal organisation may benefit by minimising danger of loss of rights that may occur when revealing the nature of the product to the potential buyer (and so run the risk of competitive innovation or misappropriation of the good), that will be a condition of sale in the external market.

Fourthly, there may be government intervention which benefits internal organisation. Where government regulations restrict external markets via price controls, the internal organisation benefits by differentiation of real and nominal prices to minimise government influence. This may be especially important where there are controls upon international capital movements or tax liabilities that aim to redistribute income within that country.

Although we see then, that in certain cases internal organisation may be favourable to the external market, Casson (1979) also distinguishes costs to internalisation that may be fourfold. The first of these may be referred to as the problem of incentive. Increases in effort or efficiency are rewarded in the external market directly, but only confer benefit to the internaliser in the internal organisation. Therefore to maintain incentive internally, the internaliser must monitor managerial efficiency and reward on the basis of measured efficiency. This tends to raise contractual costs in the internal market organisation.

A second cost may be that of market fragmentation. By fragmenting well integrated intermediate product markets, internalisation may prevent optimum scale being achieved, or possibly even minimum efficient scale. This will lead to higher costs of operation than in a well defined external market.

A third cost associated with internalisation is the problem of diversity. Internalisation will be limited by lack of diverse knowledge of production of the internaliser which prevents accurate decision making. In the international sphere with different cultures, languages, and societies this may prevent the firm from undertaking international operations.

Fourthly, there is a problem of internalisation that is exclusive to MNEs, the problem of foreign ownership. The internaliser in foreign locations may face xenophobic or nationalistic tendencies by both clients and governments and therefore faces a higher level of risk than in the external market. This nationalism will generally reflect the state of political relations between the source and host country.

Casson's analysis suggests that internalisation will predominate in two types of industry. The first is industry which relies heavily upon proprietary information. Within this type of industry internal organisation may be beneficial to:

1. avoid inadequacy of loosely assigned property rights concerning proprietary information;
2. dispose of problems relating to discriminatory pricing in the external market;
3. avoid contractual loopholes that may occur in selling information upon the external market.

By integrating R&D (which includes the development of marketing skills and management systems in this context) the firm may produce worldwide via multiplant operations, given the impact of trade barriers such as tariff and transport costs. The product of this according to Buckley and Casson will be horizontally integrated MNEs.

The second type of industry in which internalisation is predicted is where industries operate multistage production processes under increasing returns to scale or with capital intensive techniques. Internalisation is necessary in this case to allow for price discrimination, and also to internalise future markets where there are significant production lags; the more capital intensive the industry the greater will be the incentive to internalise. This is predicted to give rise to vertical integration, basing choice of location of plant operations on factor price differences and relative height of barriers to trade.

A further point that emerges from both the Buckley and Casson (1976) and Casson (1979) analysis is the importance of locational factors on the incentive to internalise. Casson has argued that:

> Internalisation will predominate in markets where there are opportunities for transfer pricing ie markets which span the boundaries of customs, fiscal, and currency areas, and will be greater the higher the tariffs, the larger the tax differentials between the fiscal areas and the more severe the restrictions on capital movements between currency areas. Internalisation will be least in markets which span areas where there are differences in social structure, culture and language, or where the countries concerned are politically hostile. (pp.61-2, 1979).

Finally, Casson suggests that fragmentation, which is a possible cost of internalisation, depends upon the nature of the product and that the market for information will generally not fragment if knowledge is proprietary and not public, because of the indivisibility of knowledge. Hence Casson's argument states that industry specific as well as firm specific, region specific, and nation specific factors are relevant in the decision to internalise. This is of course a major principle of the earlier work of Buckley and Casson.

Although this approach provides a valuable contribution to the analysis of which industries are likely to engage in overseas production and why, the predictions concerning the influence of locational factors are possibly suspect. It is argued here that major differences in social structure, culture, and language may confer an advantage upon internal organisation rather than a cost; by internalising production in that region the firm can impose structures and communications frameworks that are compatible with the internaliser and so incur lower costs than the market alternative. This suggests that the analysis of Casson (1979) has ignored the mobility of source country personnel worldwide who may operate within the confines of the firm's internal organisation regardless of social conditions. Since adaptation of corporate structure to include locational factors is likely to include a learning curve, we argue here that the more established firms in the international industry will be the leaders into difficult markets. It is further suggested that the more different the market to that of the internaliser's, and hence the greater the costs of external communication, the more likely is internalisation; thus the greater are the social differences, the greater is the incentive to internalise in that region ceteris paribus.

This criticism however does not invalidate the fundamental basis of the Buckley and Casson model, although it should be noted that the framework actually refers to two potential situations where internalisation may take place: the first involves internalisation of a market where an arm's length contractual relationship is replaced by a unified market, while the second involves internalisation of an externality to create a market where none has existed before. Casson (1984) has suggested that the introduction of locational factors into this framework leads to the argument that, given these situations in which internalisation occurs, monopolistic advantages are not necessary to explain the existence of MNEs, since benefits accruing to internalisation will be sufficient motive for FDI to take place. However, this argument can be overemphasised. Dunning (1981) for example views the existence of MNEs as a consequence of the interaction of monopolistic (or what Dunning refers to as 'ownership'), internalisation, and locational advantages. The provision of these factors within the context of firm-, industry- and country-specific characteristics provide the basis of the approach that goes some way to predicting 'where, how and why' MNEs exist. Dunning therefore argues that the presence of both ownership and internalisation advantages are necessary for the existence and continued growth of the MNE - while internalisation may be the initial impetus to FDI, the ownership of firm specific competitive advantages ensures that the enterprise can compete with other indigenous and foreign competitors in that market. As such internalisation may be part of a more general theory of the MNE.

This argument instantly weakens the view of Rugman (1981, 1982) that internalisation alone constitutes a general theory of the MNE. By dismissing ownership advantages as the response of the firm to external market imperfections,

Rugman essentially fails to recognise that firm specific and industry specific factors will be instrumental in determining the extent of overseas operations of the enterprise. A further criticism by Casson (1984) is that Rugman does not distinguish between the two possible types of internalisation, and therefore restricts his interpretation of internalisation. Rugman's claim that internalisation advantages are a product of centralised control and R&D activities seems unjustified in the context of benefits accruing from international division of the firm's activities.

2.3.2 Magee's Appropriability Theory

A similar argument to that put forward by Buckley and Casson on the theory of internalisation is that of Magee (1976, 1977) on the appropriability of MNE corporate behaviour. Magee's argument is primarily that the more sophisticated the know-how, the more likely is internalisation of that factor as a means of appropriating maximum rents to the firm. Hence firms that are engaged in R&D (in the broad sense of technical developments, marketing and management skills) to a significant degree will undertake international production in the form of internalised activities via FDI. Magee suggests that appropriability problems will occur in industries where:

1. the name of the firm is a significant advantage to production eg where products are experience goods;
2. after sales service is a necessary condition of the product that is a factor of product quality;
3. there are intrafirm complementarities between products eg the use of related information is costly to license;
4. products are new and differentiated, leading to an evaluation gap between buyer and seller;
5. there are benefits from production risk diversification and economising on errors through learning from experience in other product lines.

In each of these cases it is argued that benefits of internalised organisation are greater than in the external market.

A notable feature of the work of Magee is the dynamic structure in which the appropriability theory is based; the thesis here is that industries experience a technology cycle that parallels Vernon's (1966) cycle for individual products. The central argument is that the industry will move from compacted to expanded structure and from idiosyncratic to standardised trade in intermediate goods (technology) as a consequence of diffusion and change in product technology. Parallel works by Swedenborg (1979) and Hennart (1982) suggest a similar situation, however a criticism that has been levelled at this approach is that it is unlikely to apply to whole firms, but may be relevant only in the context of explaining parts of large corporations (Dunning, 1982). This criticism may reflect part of a more general argument against this approach that the dynamic structure of Magee's analysis does not incorporate exogenous factors that may affect the firm's corporate behaviour within the industry (eg fall in worldwide demand for the industry product as a consequence of economic recession). While this suggests that the theory is not a complete explanation of internalisation, the work of Magee does provide a valuable insight into the role of internalisation in technical diffusion, and successfully incorporates dynamic analysis into the framework. As such, Magee's work should not be overlooked as a valuable contribution to internalisation theory.

2.3.3 Diversification and Internalisation

The works of Lessard (1976, 1977, 1979) and Rugman (1976, 1977, 1979, 1981) provide a further facet of internalisation theory, by suggesting that there may be benefits of internal organisation in respect of world capital market imperfections. It is argued that these imperfections prevent individuals from enjoying internationally diversified portfolio investment, and therefore individuals undertake stockholdings in MNEs as an alternative way of international diversification.

This general idea of diversification fits into Vernon's (1983) idea that the firm may undertake FDI as a response to risk from local and foreign competitors, but there are two fundamental arguments against this approach. Firstly, financial markets in the main are among the most efficient of all markets since there are many possible transactors, information is widely disseminated, and transactions costs of market operation seem to be low.

A second argument is that the approach does not take

account of behavioural aspects of managerial control of MNEs; much of the behavioural literature on FDI stresses that it is the tastes and desires of managers rather than consideration of shareholder wealth maximisation that forms the basis of MNEs' policies. This criticism, however fails to recognise that risk averse attitudes of managers may provide a stable stream of earnings that is consistent with the approach.

In view of the criticisms of the approach, it is suggested here that to explain the existence of MNEs on the basis of financial market imperfections is to overstate the case; however, the fact that sovereign and institutional impediments to capital do exist, and the limited validity of the second criticism above implies that the approach should not be dismissed out of hand, and may be added as a potential factor favouring international production.

2.3.4 Internalisation and oligopolistic strategy

The basic argument put forward by Casson (1985a) is that the oligopolistic strategy of MNE activity provided by Graham (1974, 1978) may be extended and explained in terms of internalisation theory. Casson extends Graham's analysis to suggest that Knickerbocker's (1973) model concerning oligopolistic dependence of MNEs might be linked to Graham's. The 'follow the leader' policy may be taken as a case of rivalrous FDI by company A in company B's major overseas market (given that B is of the same nationality as A) to warn off any action within the home market by B that may affect A's domestic market position. A similar argument may be that A's rivalrous FDI in B's overseas market is designed to spoil that market for B to prevent B from exploiting any advantages attributable to that overseas market that could effectively be reflected in B's market position.

Casson argues that rivalrous FDI will be undertaken as a defensive move, and will take place within the internal organisation rather than the external market for two reasons:

1. Rationalisation of the aggressor and defender policies in the defender's home market would involve a sophisticated agreement between the firms that would be prohibitively costly for them to enforce, and would carry high probability of opportunistic behaviour on the part of

the aggressor. The failure to agree on rationalisation will ultimately lead to the threat of rivalry behaviour on the part of the defender.

2. For the threat to be carried out efficiently, the defender must coordinate activities in the home and host market to pose a serious threat. By coordinating activities using the external market, the defender runs the risk of losing control of pricing strategy in the host country (say to a licensee), which in this case would leave the defender with no bargaining tool with which to threaten the aggressor in the defender's home market. Internal organisation is likely to be the least costly and most effective way of maintaining a coordinated strategy against the aggressor in both the home and host country markets.

Casson therefore suggests that in the case of rivalrous behaviour, FDI provides a more effective instrument than use of the external market on the basis of advantages of the internal organisation. As with the general theory of internalisation, this occurs because of market failure or imperfection.

In a similar vein to Casson, Vernon (1983) has argued that the internal organisation will be the preferred mode of operation in situations where this reduces the element of risk faced by the company. Vernon suggests that internal organisation will therefore occur most when there are few firms in the market, the competitive environment is uncertain, and collusion is difficult to enforce. Risk is likely to be raised when this type of situation occurs in a multinational context.

Vernon emphasises that in an international oligopoly the internal organisation will be the preferred mode of operating as a means of reducing uncertainty that will occur through 'follow the leader' and 'rivalry behaviour' type oligopolistic strategies. This will particularly be the case in globally oriented type oligopolies where competitors consider strategy on a worldwide basis rather than in terms of Graham's initial idea of home and host country retaliation, and exchange threat and counter threat in a situation of differing (or perceived to be differing) market share through aggressive action. Vernon argues that in this situation risks to all oligopolists involved are such that they have to rely upon internal organisation for worldwide coordination of strategy that could not effectively be undertaken in the external market because of the potential costs of policing an agreement, default on policy by a licensee, or opportunistic behaviour on the part of the licensee that could benefit a competitor. For these reasons strategic moves by oligopolists in global competition will undertake FDI rather than use the external market to maintain market share through strategy along the lines of Graham or Knickerbocker. Casson and Vernon follow similar lines by suggesting that the 'follow the leader' and 'exchange of threats' may both be part of a wider overall strategy on the part of the oligopoly.

It is perhaps worth noting within the context of these works by Casson and Vernon that the subject of internalisation and oligopoly strategy does not provide a complete theory of FDI, but rather an extension of the existing body of theory; as Casson suggests, the notion is not a complete theory since the framework fails to provide reasons for the initial investment which may come from 'technology transfer, or some other motive' (1985a, p.34), that is the impetus for the aggressor's move into new markets. A more serious criticism however is that Casson suggests that the defensive oligopoly will be motivated by technological weakness on the part of the defender (whose market is entered by the technologically stronger aggressor who aims to increase market share) and therefore it is the lack of transferable advantages on the part of the defender that causes the defensive investment. While it may be true that the defender is weaker than the aggressor in terms of advantages, it is highly unlikely that the threat of the defender will have any impact on the aggressor's home market without a transferable advantage. As such it is argued here that while Casson's extension of the theory provides useful explanations of internalisation motives and the strength of competitive advantages of the aggressor and defender, the analysis concerning the ownership of transferable advantages overstates the case against the defender and therefore fails to reconcile all of the criticisms levelled at Graham's model in the previous section.

2.3.5 Transaction costs within internalisation

While the concept of internalisation provides a key element in the theory of the MNE, at its most general and without additional assumptions it is almost tautological. To render this theory operational it is useful to specify assumptions about transaction costs. Transaction costs refer to the costs of attempting to overcome obstacles to trade and therefore relate to the choice between internal and external markets. Given that transaction costs exist, the profit maximising rational transactor will choose the arrangement that experiences the lowest cost (Casson, 1982).

Teece (1983) has argued that given that firms often possess intangible assets (eg managerial and/or marketing know-how, patented designs) the firm will internalise the asset if:

1. The asset in question is not fully exploited ex ante. External use of the asset may not occur where there are transactional difficulties arising from recognition, disclosure and transfer costs of the asset (Teece 1981, 1982). This is most likely in the case where the asset is tacit knowledge embodied in the firm's product (Williamson, 1975, Buckley and Casson, 1976).

2. Arm's length transactions in the servicing of the intangible asset are exposed to higher transactions costs than internal use of the asset. Both Casson (1982) and Teece (1976, 1983) identify a number of possible market failures that may lead to internalisation because of the inability of unassisted markets to protect and service quality.

This notion of quality protection is central to the approach of transaction costs as an internalisation motive. Casson suggests that 'psychic distance' - a function of social culture differences and distance from home country - raises transaction costs by impeding information flow between traders, and that increase of communication of product quality (either first or second hand) enhances the firm's standing and overcomes psychic barriers. A corollary of this is that the reputation of the person or product will be a major factor in information flow which may impede or benefit trade, and if beneficial will reduce the relative cost of arm's length trade.

Casson has further argued that the onus of quality control rests with the producer because of economic efficiency arguments. This is based upon the idea that the producer has direct access to production, fully understands the production process, and stands to lose most if quality of the product is defective. In cases where reputation is a significant factor in transactions:

> Economic efficiency in quality control ... normally requires that the producer should assume responsibility

for quality control, and that people buying from the producer should have confidence in the producer's quality control. (1982, p.34)

It should be noted that as far as customers are concerned a reputation for product quality may be the basis for a monopolistic advantage, which is probably most important where a brand name is used to indicate quality.

If this is the case, licensing or hiring out of this advantage in various regional markets will most likely not occur where the commodity sold under the name is an intimate combination of the physical product and image and reputation of the producer. In complex or differentiated products it will often be impossible to distinguish product from image or reputation, and therefore any inferior product manufactured by a licensee would have unfavourable repercussions upon the licensor's reputation either personally or through the production process and product. This analysis leads us to the major predictive element of the transaction costs approach - the greater the potential costs of underperformance of the licensee to the licensor, and as a consequence the greater the costs of insuring against underperformance (including legal constraints and guarantees concerning usage of the asset, monitoring of licensee product, and search costs of finding acceptable licensees), the more likely it is that that asset will be exploited internally rather than licensed through external contracts.

The implication of this approach is that the concept of internalisation may be extended beyond explaining MNEs in R&D intensive industries, to explain other forms of multinational involvement that may be based upon factors other than proprietary knowledge. The transactions cost approach explains the MNE which specialises in making markets in consumer products, and as Casson suggests, this extension of the theory gives internalisation greater explanatory power in several types of markets:

> Consequently there are very few markets where the presence of market making MNEs can prima facie be ruled out. The theory predicts, however, that in whatever markets MNEs occur they will be most strongly represented in those segments of the market where quality is at a premium. (1982, p.38)

It should be noted, however, that this theory also predicts where internalisation will not take place; where the product is easily definable and quality control is enforceable without effort the transactions costs may well be less than internal benefits, and licensing may take place. Hence it is argued that while this approach is an extension of the theory of the MNE, it is also conversely a theory of the uninational firm, and may be useful in determining how markets are serviced given characteristics of the product of the firm. This subject is treated in greater detail below.

2.3.6 **Modal choice of market servicing**

Much of the literature on the MNE fails to recognise that markets may be serviced in one of three general ways; export of product from home country of the firm; licensing of host country producer to make the firm's product in that market; or FDI to undertake direct production in that host country. The choice between exporting and FDI was initially taken up by Hirsch (1976) who concentrated upon the comparison of export marketing cost differential against the additional cost of direct foreign operations, and argued that the least cost option would be taken up given that certain locational factors were present. This idea of export/FDI option was subsequently extended by Lall (1980) who has argued that transferability of monopolistic advantages is the crucial element in this choice - where transferable advantages exist, the firm will undertake FDI to benefit from utilisation of the advantages in the foreign market. Within this argument non-transferable advantages clearly lead to export.

While these approaches are useful in connecting ownership advantages with locational factors, both works may be criticised for not including licensing as an option of market servicing (ie in more general terms, the choice between the external and internal market). Buckley and Davies (1979) incorporate licensing into this analysis within an internalisation type framework - if internal markets are least cost then locational factors will determine whether exporting or FDI are undertaken, whereas if the external market is least cost licensing will ensue, given locational factors. Buckley and Davies have suggested, on this basis, that licensing will predominate:

A Literature Review of MNE Theory

1. Where advantages are identifiable and transferable at low cost.
2. Where indigenous oligopolistic competition in a host market could have unfavourable oligopolistic reactions to new entrants to the market.
3. Where the firm or industry is constrained by exogenous factors (eg lack of corporate finance, host government protectionist policies).

The issue of market servicing mode has been taken up in a dynamic sense by Rugman (1980) and Buckley and Casson (1981). In both cases, fixed and variable costs of possible operations provide the basis of choice of mode of market servicing. Buckley and Casson argue that for a simple product (ie relatively straightforward in all modes) growth of market demand will lead the firm to move from exporting to licensing and finally FDI in an effort to minimise costs. Rugman's model is similar, although he incorporates diffusion of technology into the model, and therefore moves from exporting to FDI and finally to licensing as diffusion of technology increases. In that Rugman's model may be criticised for failing to allow for continual product development, and also assuming that there will be an external market for technological factors already widely diffused (since it is unlikely that a potential licensee will pay for partially diffused technology) it is argued here that the Buckley and Casson model provides the most complete dynamic framework of modal choice of the MNE. This is further illustrated by the fact that Rugman's approach is limited in that only one component of potential advantage is catered for, whereas the Buckley and Casson model provides a more general framework of MNE modal choice.

2.3.7 Other forms of Foreign Involvement

Although we have assumed that the firm faces three possible modes of market servicing, it is possible to include non-equity forms of foreign involvement within this analysis. The most common forms of these are management contract and joint venture, and it is argued here that the existence of these types of foreign involvement may be explained with reference to internalisation theory. The primary factor that enables and encourages non-equity forms of foreign involvement is the ability of the firm to control its rent yielding assets. Dunning (1982) argues that if assets or advantages owned by the firm are protected from misuse or diffusion, then management contract or joint venture may take place, depending upon the circumstances:

1. Management contract may be a feasible option in a market where the firm can both fully appropriate the economic rent of the asset in question, and where provisions for protection against dissipation or misuse of techniques or information are satisfactorily catered for in the contract;

2. Joint venture may be undertaken where intangible assets are not easily externalised or protected by contract but may be used by the owning company without fear of imitation or adaptation because property rights are clearly assigned (eg name of the firm as a legally defined enterprise). In this case the advantage may be used by the firm in conjunction with other enterprises' production schedules by the firm to earn the economic rent attributable to that asset, without fear of diffusion of the asset through an industry development cycle as outlined by Magee (1977).

If these conditions do prevail, non-equity involvement is likely to occur where there are benefits of this type of arrangement. In the case of management contract, the firm may face lower fixed costs than in FDI, and therefore face less risk exposure. A similar argument can be put forward for joint venturing (Vernon, 1983) although an additional benefit attributable to joint venturing is that of potential economies of joint production and marketing that will be a feature of this type of foreign involvement (Dunning, 1982).

2.4 LOCATION THEORY AND THE MNE

In contrast to the industrial organisation approach and internalisation theory, location theory elements of the MNE have been given little direct coverage. In the light of the argument that location theory not only provides an answer to 'where' international production takes place, but also answers why the firm goes abroad at all, this is a serious shortcoming. Vernon (1974) has incorporated location factors into the PC hypothesis to illustrate the interplay of dynamic industry development and locational influences, but apart from this much of the literature on the theory of the MNE tends to implicitly include locational factors as a minor consideration. The importance of location theory to the MNE has been taken up by Buckley (1985), who suggests:

A Literature Review of MNE Theory

> Location factors ... enter the theory not only in their own right, as an influence on the relative costs facing an MNE with a choice of locations, but also may provide motives for international expansion. (1985, p.13-14)

It is argued here that location factors may be analysed on two fronts. The first one is the characteristics of location that influence FDI, ie why and where does the MNE go abroad? The second relates to advantages attributable to producing in more than one location.

2.4.1 Choice of market of the firm

Because location specific factors are external to the firm and often immobile, the study of location factors and MNE activity suggest that, ceteris paribus, if the firm can benefit more from production overseas than production in the home country because of location specific factors, the firm will locate in that market. Buckley (1985) has suggested that there are three location specific endowments of particular importance to the MNE:

1. raw materials, leading to vertical FDI;
2. cheap labour, leading to 'offshore production' facilities;
3. protected or fragmented markets, leading to FDI as the preferred means of market servicing.

In terms of economic argument, possibly the most useful tool for explaining the motivation for producing overseas in this context is that of comparative advantage (Buckley and Casson, 1976). The notion that MNEs search abroad for the cheapest production facilities on the basis of their need for cheap capital or labour forms the basis of locational influence in the Vernon PC model (1974), but this idea is also implicit in the analysis of Hirsch (1976) and mentioned as a relevant factor in other works into MNE activity (eg Dunning, 1981).

The heterogeneity of location factors has been instrumental in explaining not only the choice between home and host country for production, but also the criteria behind choice of overseas site of production between one or more locations. Possibly the most common locational determinant referred to is that of potential market size. The potential growth of the foreign market has been included as a major locational factor in several analyses of the MNE (Knickerbocker 1973, Casson and Norman 1982, Vernon 1974, Rugman 1980, Buckley and Casson 1981), but is only one of many potential locational attractions. These include barriers to trade in the form of tariff walls and the like (Buckley and Casson 1976, Casson 1979), ability to exploit economies of scale in the market (Vernon 1974, Knickerbocker 1973), social and cultural affinity to the MNEs home country (Casson and Norman 1982, Casson 1979, Dunning and Norman 1979), market stability (Knickerbocker 1973) and ability of the indigenous competition to maintain barriers to entry (Graham 1978, Buckley and Davies 1979).

While these factors may be considered of particular relevance to the industry analysed (eg Dunning and Norman 1979 attempted to analyse location factors influencing MNE office location) it is not necessarily the case that they sum up all relevant location factors for all industries. The movement of US banks abroad to maintain US MNEs clients' accounts in those markets for example suggests a highly relevant industry specific factor that may not be relevant elsewhere (Yannopoulos, 1983). Hence it is argued here that it is the nature and structure of the industry and firm (that will be affected by the product of the industry) that will determine what locational factors motivate overseas production, and that these will be acted upon only where there is a distinct advantage for the firm to undertake FDI rather than exploit those factors in some other way. Ultimately this may mean the interaction of ownership advantages and location factors that may affect choice of location, which is discussed in more detail below.

2.4.2 Factors affecting multinationality

A second factor that is both a consequence and motivation of overseas production is factors that occur through the firm being multinational. The argument here is that the firm faces a different situation to unilateral firms because of locating in more than one market, and therefore modifies policy and behaviour to the unique situation that occurs because of multinationality.

Although there may be both benefits and costs of multinationality, the benefits have been emphasised most in the literature, and of these transfer pricing has received the most attention. The ability of the MNE to exploit national tax differentials has been regarded by many economists as one of the major motivations behind MNE activity (for example Casson, 1979) but may be considered part of a greater benefit accruing to MNEs as a consequence of host government differential policies on many aspects that influence production (Buckley and Davies, 1979). Other benefits of multinationality refer to differentials already outlined above, and reflect the ability of the firm to exploit its ownership advantages within an international framework. Dunning (1981) has suggested that there are three advantages related directly to multinationality:

1. Multinationality enhances ownership advantages of the firm by offering wider opportunities for exploitation;
2. Multinationality allows more favoured access and/or better knowledge about inputs, information, and markets;
3. Multinationality increases the ability to take advantage of international differences in factor endowments. The ability to diversify risk is also increased eg via different currency areas, and exploitation of capitalisation ratio differentials.

On the subject of diversification Vernon (1983) has argued that FDI may be undertaken as a response to risk to stabilise demand for the firm's product and ensure efficient production (which is an adaptation of internalisation theory), but also suggests that there are potential costs associated with multinational production. Of these, exposure to risk of loss of fixed assets are possibly the greatest, although discrimination against the MNE for nationalistic or xenophobic reasons may also be a significant cost (as initially argued by Hymer, 1976). The firm may take measures to alleviate these costs (Vernon 1983 for example suggests that joint venture may provide acceptable diversification) but may only be practical in certain instances; where the potential benefits are outweighed by the costs of multinationality, MNE activity is likely to be minimal.

2.4.3 Kojima's comparative advantage theory

The approach put forward by Kojima (1978) is based upon

relative comparative advantage of both the source and recipient country of FDI, and therefore is based upon exploitation of locational factors as the motivation for FDI. Using earlier work by MacDougall (1960) and Kemp (1962) into the welfare aspects of foreign investment for a host country as impetus, Kojima distinguishes between trade oriented (Japanese type) foreign investment and non-trade (American type) foreign investment. The principle of the approach is that the Japanese type investment takes place in response to relative dynamic comparative advantage changes, and corresponds to a macro economic government policy. This consists of locating declining home country industry that has lost its comparative advantage in regions where competitiveness may be regained because the comparative advantage is relatively more favourable. In terms of Japanese FDI, Kojima suggests that the 'en masse' movement of Japanese firms in specific industries to other countries has taken place to exploit natural resources not available within Japan, and also to switch labour intensive activities from high cost to low labour cost locations. This has inevitably led to movement of Japanese firms to LDCs in particular, and is supposedly trade oriented because it increases technology transfer to these developing countries, and also promotes two way trade between Japan and the host countries where Japanese firms are located. Thus Japan gains minerals and raw materials necessary for production while the host countries gain managerial skill and technology for indigenous production.

The American type FDI, on the other hand, is based upon microeconomic principles in Kojima's model; the motivation behind this type of investment is defence of oligopolistic position in world markets, exploitation of factor markets, and response to trade barriers. Such investments are accordingly anti-trade in the Kojima model because they transfer activities from where they have a comparative advantage to where they are disadvantaged, and therefore reduce trade especially between developed and developing countries since they do not take account of trade creation which will be a consequence of production that embodies relative comparative advantage.

Ozawa (1979) has extended the Kojima model to take the distinction between Japanese type and American type FDI even further. Ozawa suggests that in the Japanese type FDI there is no need for monopolistic advantages because the macroeconomic approach to FDI relies upon the exploitation of industry specific rather than firm specific knowledge. Hence in this type of FDI there is no motivation for oligopolistic reaction FDI as presented by Vernon (1971) and Knickerbocker (1973) which is the rationale of the American type FDI that is trade inhibiting.

The work of Kojima and Ozawa on the Japanese model of FDI provides us with a clear picture of how locational factors may be instrumental in FDI, and is also original in that the approach is based upon macroeconomic analysis. However, these points do not lessen the importance of the serious defects in the analysis which raise questions about the validity of the model:

1. Kojima's notion of trade inhibition and trade creation, which provide the basis for the two distinct types of FDI, are not clearly discernible in the context of Kojima's argument. Kojima fails to recognise that even within the American type FDI which he describes, the recipient country may benefit from the creation of comparative advantage as a direct consequence of FDI. In this case, American type FDI is therefore not necessarily trade inhibiting, which clearly limits the argument put forward in the model.

2. Kojima argues that Japanese industry can easily be transferred to LDCs because of the small technological gap between the two. As Japan has experienced rapid industrialisation this gap has grown to the point where now Japanese type FDI may be similar to American type FDI. Hence Kojima can be criticised for failing to recognise his static analysis as part of a dynamic pattern of growth of Japanese MNEs.

3. Following on from the point above, Ozawa's notion of Japanese MNEs relying on industry specific advantages is redundant if Japanese MNEs enter developed country markets where indigenous competition is firmly established. Since Japanese MNEs now seem to be following this pattern of FDI, firm specific advantages are essential to effective competition, which brings Japanese MNEs more in line with the American type MNEs of the Kojima-Ozawa analysis in terms of competition.

4. A major criticism of the Kojima model put forward by Dunning (1985) is that Kojima fails to take account of benefits arising from the internalisation of intermediate product markets; where Kojima emphasises internalisation he regards it as inferior to the market situation and generally related to firm specific monopolistic strategy.

This argument may be regarded as a consequence of Kojima's adoption of a neoclassical perfect framework in his model, that does not allow for market failure of any kind. It is suggested here that in consideration of the importance of internalisation theory as a powerful tool for the explanation of MNE activity, this framework is wholly unsuitable for this type of analysis, and casts doubts upon the validity of Kojima's 'trade creating' and 'trade inhibiting' notions.

2.5 GENERAL THEORIES OF THE MNE

As a consequence of the failure of the approaches outlined above to provide a complete picture of the MNE, there have been several attempts at more general theories of the MNE based upon extensions of existing literature and approaches.

It has already been suggested that Rugman (1980, 1981) has claimed that internalisation alone represents the general theory of the MNE. In this model, Rugman suggests that ownership and locational factors as influences on the MNE are both consequences and a part of the internalisation process. Without reiterating the criticisms of Rugman's approach already outlined, it can be argued that this approach is invalid because Rugman fails to account for the interaction of ownership, internalisation, and locational factors, and therefore the framework is at its most general, tautological. In addition Rugman limits his interpretation of internalisation and often confuses the issue, and therefore his general theory of the MNE may be disregarded as invalid.

Calvet (1981) has put forward the 'markets and hierarchies' approach associated with Williamson (1975) as a candidate for a general theory of the MNE. Calvet argues that although international involvement involves aspects of location theory and industrial organisation theory, the transition from market to MNE hierarchy may be explained by minimisation of transactions costs that occur because of market imperfections (Casson 1982). While this approach appears more systematic than that of Rugman, the notion of bounded rationality which forms a basis for adoption of the hierarchy may be dismissed as uncharacteristic of the neoclassical framework on which the theory is couched (Buckley, 1985). This leads on to a third attempt at a general theory which is similar to Calvet's approach, but possibly more general in that the origins of the theory lie in industrial organisation, internalisation and location theory synthesis. This eclectic approach put forward by Dunning (1977, 1979, 1980, 1981, 1985) suggests that necessary and sufficient conditions for FDI are that there are ownership (O), internalisation (I), and locational (L) factors that may be exploited by the firm undertaking international production. Of major importance to the theory is the interaction and interdependence of these three conditions, so that the exclusion of one type of advantage will lead to foreign markets being serviced by some form other than FDI, or possibly not considered at all.

It is argued here that Dunning's approach to internalisation is favourable to that of Calvet's because of the interaction of O and I factors that is an essential part of the framework. Dunning (1985) suggests that O advantages may occur either through structural market imperfections (as emphasised by Hymer, 1976 and variants on the industrial organisation approach) or transactional market imperfections (as suggested by Casson, 1982), the type of market imperfection depending upon industry and firm characteristics. The consequence of this is that Dunning's O advantage is not necessarily of monopolistic character, but may include any advantage that the MNE possesses over indigenous firms, including advantages accruing both to internalisation and to factors of multinationality that could realistically be put down either to internalisation of several markets or locational factors (as suggested in the preceding section). In this sense it has been argued that Dunning's inclusion of both O and I advantages as necessary conditions for FDI is almost tautological. However this argument may be construed as invalid if it is considered that O and I factors reflect both the ability of firms to internalise factors which give MNEs a competitive edge over other MNEs and indigenous firms, and the choice between internal and external markets as a basis for internalisation given market conditions. As such O and I factors analyse different facets of the same situation, but are both instrumental in sustained profitability and growth of the MNE in foreign markets.

A second criticism of Dunning's eclectic theory is that the framework does not allow for dynamic predictions concerning the interaction of OLI factors. Dunning has attempted to dismiss this criticism by the adoption of an investment development cycle (Chapter 5, 1981 & 1985) which explains a country's propensity to engage in outward

A Literature Review of MNE Theory

FDI according to:

1. stage of economic development;
2. structure of domestic factor endowments;
3. degree of transaction market failure of intermediate products internationally.

Basically in the model, the country will discontinue as an inward recipient of FDI and become an outward investor as the home country economy develops, given that transaction market failures exist and domestic factor endowments do not satiate domestic demand for production. Hence as the country's income grows, there will be a dynamic progression of industry in that the country will replace inward FDI with outward FDI enabling home country firms to ultimately become outward investors, as O advantages are best exploited in a foreign location rather than via home production and export.

While this approach is attractive in that Dunning's correlation of FDI with country income seems justified to a certain extent (Chapter 5, 1981) there are two basic flaws in the framework:

1. The analysis involves a macro oriented application of micro derived variables that influence firm behaviour and therefore simplifies the dynamic macro process of FDI beyond realistic limits.
2. The dynamic nature of the approach is based upon static variables and does not account for exogenous variables that may affect the situation through time.

The fact that these criticisms are directed at an extension of the eclectic theory is relevant to the fact that the eclectic approach remains valid despite these criticisms, though clearly limited to static analysis if the arguments above are justified.

In terms of a general theory, the eclectic approach seems to be the most well defined and generally applicable to the involvement of MNEs that is available. The incorporation of industrial organisation, internalisation, and location aspects of MNE involvement, and the interdependence of these factors within the model, provides a flexible basis upon which theoretical and empirical study may be directed in that various aspects of MNE involvement may be scrutinised under the auspices of the OLI framework. A point of reference to the eclectic theory should perhaps be mentioned here; although OLI factors may be classed separately, the interdependence of these components is an essential part of the framework that suggests that clearly defined limits between the OLI advantages may not exist. This remains a reflection of the complexity of factors affecting MNE involvement rather than a criticism of the eclectic approach.

Chapter Three

AN ECONOMIC OVERVIEW OF THE CONSTRUCTION INDUSTRY

One of the major hypotheses upon which this thesis is based is that although the international contracting industry has characteristics that may be found separately in other industries, the specific combination of these factors is unique to the industry. To identify the nature and influence of these factors requires an understanding of the components that make up the industry, through the pre-construction, construction and post-construction phases. This is summarised in Figure 3.1.

Figure 3.1: The Construction Process

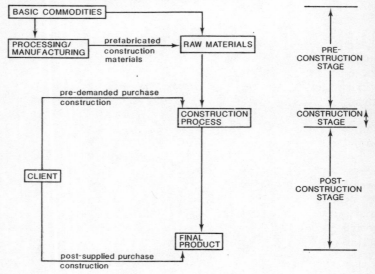

The figure suggests that the construction process is intermediate in that it involves the transformation of raw materials to the final product (eg bridge, house, dam). The acquisition of the final product by the purchaser (herein called the client) may take place either before construction has been started (pre-demand purchase) or after construction has begun (post-supply purchase). In that each is determined by the nature of the product, it is useful initially to discuss the role of the client in the construction process.

3.1 THE CLIENT IN CONSTRUCTION

The client will only indirectly demand construction activity as a means to obtain the final product, which will be used to create additional goods and/or services. Given that the final product is therefore an investment good, client demand is likely to relate to one of four types of demand for investment goods:

1. as a means to further production of goods and services, eg factories, offices, and industrialised buildings;
2. as an addition or improvement of the infrastructure of the economy, eg roads, power stations;
3. as social investment, eg hospitals;
4. as an investment good for direct utility, eg housing.

Because construction activity may be demanded for these purposes, in general clients will be one of three types; public sector, private sector, and in house private or public sector. The public client (ie central or local government) will be both a producer and provider of economic and social investment, and therefore may demand all four types of investment good. The private sector client is likely to demand means to directly increase production of goods and/or services or utility, and therefore will concentrate on 1 and 4 in the list above. The in house client (that is defined here as owning the means to construction, typically in a corporate structure) will generally be motivated by the need to increase production only (and may for the sake of this analysis be regarded as a private client). Exceptions to these groupings are likely to exist, but will generally be infrequent and therefore are not considered relevant in this general context.

It is possible to further separate clients of construction activity into those who are involved in pre-demand purchase and those who require post-supply purchase (cf Figure 3.1). Ultimately the provision of these types of final product will depend upon demand and supply of the final product. Pre-supply purchase and construction will generally take place where the contractor is unwilling to speculate on future demand of a type of product because it requires idiosyncratic development and construction related to the client to produce a specific final product at a required site. Supply of this type of final product before demand will therefore be minimal so that the client cannot buy the product 'off the peg'. Post-supply demand, on the other hand, for goods that may be built speculatively (ie on the speculation that demand for the product will exist) will accord with high demand general purpose construction that only needs locating within a general area. In this case supply of the final product before or during demand is likely to exist without the need for substantial modification for the individual client and may therefore involve post-construction purchase.

Within this analysis, it is predicted that the contractor is likely to get involved with non-speculative (pre-demand) construction where future demand is unpredictable, which may be the case in civil engineering projects requiring specialist services at a specific site (eg bridge, dam). Speculative construction is obviously more likely in predictable demand situations or where demand may be motivated, and therefore may be apparent in house, office, and general purpose factory building.

Table 3.1 summarises the relationship between the three types of clients and speculative and non-speculative construction. Speculative construction is unlikely to be common within the public sector of the economy for institutional reasons and the need to budget for and plan development according to economic and social needs, it is also not applicable in the in house case where it would be an inefficient way of coordinating between the owner and the contractor. In the private sector, however, speculative construction may be a major aspect of demand because of the nature of the products that may be constructed in this fashion (ie certain factories and housing as discussed above). Non-speculative construction is likely where coordination between the client and contractor is required, and therefore will be required in all three sectors.

Table 3.1: Client Demands for Construction Works

Type of Client/ Type of Construction	Speculative (Post-supply purchase)	Non-Speculative Pre-demand purchase)
Public Sector	Minimal	Yes
Private Sector	Yes	Yes
In House Client	No	Yes

In that speculative construction requires the production, marketing and then selling of the product, this sector of the industry is of little interest to this research because of the similarity to manufacturing industry. What is of more importance is the non-speculative pre-demand type construction, which is clearly differentiable from manufacturing in several respects that may be considered a function of special characteristics of the industry. In terms of the client, two major factors differentiate this type of construction from the manufacturing 'norm':

1. The client pre-demands the final product in order to obtain a custom built final product at the site required by the client;
2. The client purchases the final product before it is manufactured.

Given these characteristics, and the fact that pre-demand construction may be common for both public and private sector clients, this analysis is restricted to pre-demand construction, that it is argued is more in line with the objectives of this research than the investigation of speculative type contracting.

3.2 RAW MATERIALS

Within the context of this chapter, it is important to notice that raw materials as illustrated as the pre-construction stage in Figure 3.1 will generally not be manufactured within the contracting process, but will be products of other industries. The needs for raw materials by the contractor to construct the final product utilising these materials is likely to be diverse, so that products required will range from the unprocessed (eg sand) to more complex manufactured articles, such as structural steelwork; however, in that these products will combine to form the final product when assembled through the construction process, all are considered here to be the raw materials of the industry within the scheme of production as illustrated in Figure 3.1.

3.3 THE CONSTRUCTION PROCESS

The construction process as depicted in Figure 3.1 refers to the transformation of raw materials to the final product, which is carried out by the construction contractor. In this section two major features of the industry are analysed that influence and characterise the structure of the construction process in general. These are the nature of the product of the contractor, and the method of selection of contractor by the client.

3.3.1 **The Product of the Contractor**

In the sense that the final product is a combination of products manufactured by other industries, the intermediate product of the contractor in the industry is basically assembly and management of these raw materials to form the composite work (ie the final product). It is therefore the case that the intermediate product of the contractor is the provision of services in order to construct a given product. As such the product that the contractor offers will rely upon human rather than physical capital which may take several forms, and ultimately will come down to technical and managerial competence and also experience. On this basis it is argued here that supply and demand within the industry will revolve around the provision of these factors. As the client ultimately requires the contractor for intermediate use only (ie the construction of the final product), demand is likely to be based upon the level of skill of the contractor and this will accordingly influence supply. Harrison (1982) for example has argued that the contractor should be strongly influenced in his choice of bids by 'the abilities, experience, and inclination' that the firm possesses. In a similar vein it is argued here that the expertise that the contractor can offer will be the limiting factor in the supply of contractors for any one bid. A building that requires a high degree of technical skill can only be constructed by a firm with knowledge and expertise in that field, and therefore expertise of the contractor will limit the supply of firms in any bid situation where anything but basic construction is required. In this sense, expertise of the contractor clearly affects both demand and supply in the construction industry.

The fact that expertise forms the basis of the contractor's product has led to two factors that may be associated with the industry:

1. Entry into and exit from the industry are not likely to be restricted by barriers that are prevalent in many oligopolistic manufacturing industries. Because construction involves the provision of services in the form of human rather than physical capital, any person who has the know-how concerning a construction skill has the potential to enter the industry as a provider of that service. The high turnover of entrants and leavers in the industry at any one time may be put down to two factors:

i. At the less skilled end of the industry, expertise and know-how is not difficult to obtain so that entry to the industry is relatively unrestricted. However it should also be noted that not all skills will be so easily learnt or readily accessible so that entry to specialist sub industries may be limited.

ii. The provision of working capital is often catered for by the client so that the contractor needs little capital to enter the market and compete; the client generally pays initially for materials, equipment, and other factors necessary for construction so that the contractor need not have extensive capital reserves to undertake a project, but may rely upon skill needed to construct the final product. This may even be the case in speculative construction if the developer and contractor are not the same person, though clearly where the contractor does finance speculative construction the provision of finance is the contractor's responsibility.

An implication of the ease of entry and the need for low capital requirement is that within the industry size of firm may range from a one man operation to a large company, since each may find a demand for the product they offer. While this may be considered a function of ease of entry and the nature of the contractor's product, characteristics of the final product may also be instrumental in this, as discussed later in this chapter.

2. Although any type of construction activity comes

under the auspices of the construction industry, in reality service and management differentials according to the technical skills necessary imply that there are many sub industries within the industry as a whole, ie the industry is highly fragmented. The incidence of fragmentation within the industry is likely to be high for three possible reasons:

    a. The final product is an investment good that is itself required for further production. The contractor therefore must have skills in relation to these needs that are likely to be highly diverse according to client demands.

    b. Knowledge may not be readily transferable between different types of construction without some cost (in either time lost acquiring the expertise or buying in the necessary skill); the contractor is therefore likely to specialise in areas in which demand for the firm is high, and sub contract or not enter other areas that require additional skills.

It is therefore likely that the industry as a whole will be fragmented according to the expertise and skills associated with different aspects of construction.

In the sense that skill is transferable between contractors, size of the firm may be related to ability to enter more than one of these sub markets, which may be considered a consequence of the ease of entry to the production process. Given that the firm has adequate financial resources, the ability of the firm to buy in expertise in management within a particular area of the market is a factor attributable to the ease with which this expertise may be transferred.

Figure 3.2 illustrates the dynamic situation in which fragmentation in a specific sub industry is likely to occur. In general there are three separate stages of the construction process, in a vertical relationship, that require different types of expertise. The three distinct components that make up the construction process are initial design and specification of the product for the client, construction of the product based upon the designs, and the sub contracting of specific parts of construction that the major contractor cannot carry out. All three will additionally involve some input (either information about materials or supplies of inputs for the construction process) that form the basis of the final product. The analysis behind Figure 3.2 is intuitive; the client will usually not have sufficient in-house

An Economic Overview of the Construction Industry

Figure 3.2: Construction as a Vertical Relationship

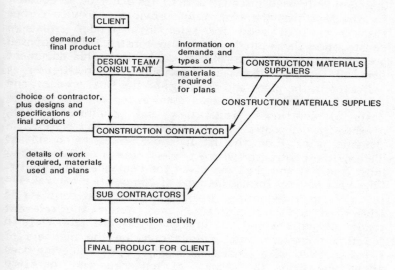

knowledge to plan the final product and select the appropriate contractor for the job, and therefore will generally appoint professional advisors for the task, who may be civil engineers, architects, or in some cases quantity surveyors. The appointed consultant will then draw up plans of the final product based upon client demands and materials available, and choose the contractor for construction of the product using a basis of selection which may involve several factors about the contractor in addition to the estimated cost of the final product. The contractor considered most able to carry out the work by the consultant will then undertake the contract under the supervision of the consultant according to the original plans. Where the contractor does not have expertise or skill required for a specific part of a project, that part will be sub contracted to another firm who has the required skill, so that even in a specific sub-industry fragmentation may occur.

3.3.2 **Selection of Contractor**

Given the nature of the product of the contractor, it is

argued here that the adoption of the bid and tender system (as commonly used in construction) has occurred to account for the fact that the contractor provides a service rather than a physical good for sale. In a tender situation, construction firms put in a price to construct the proposed project based upon the 'bill of quantities'. This document lists all material items used within the construction as indicated in the plans and drawings of the final product, and the contractor then prices each item in the bill and includes costs of additional preparatory and preliminary work, overheads, and profit margin necessary for the project, in the final price. From the list of tender prices the consultant chooses the contractor whose tender price is considered the most realistic and cheapest given the demands of the client. There are several forms this type of competition can take:

1. Open tendering. In open tender, the contract is advertised and any contractor may obtain the relevant tender documents and put in a price for the job. While in practice there are usually limitations on the numbers of firms supplied with tender documents, open tendering does allow new firms the chance of tendering and therefore remains a safeguard for competition within the industry.

2. Selective tendering. In the selective tender only a limited number of selected contractors are invited to tender based upon some criterion of the client/consultant that has determined those contractors most able for the project. The criteria often required are managerial and technical capability, experience of the type of project contemplated and financial ability (Ashworth, 1983).

3. Two stage tendering. In this case there is a pre-tender which may be selective or open, where the contractor puts in a basic tender price. This is followed by a second stage where the most promising contractors (on the basis of the pre-tender) are invited to submit a second tender that is considered to be a realistic bid for the work. This may be followed by negotiation of the second phase bid price between the contractor and client to reach a mutually acceptable situation of work and price.

4. Negotiation. Negotiation may involve both pre-selection and post-selection phases; in the pre-selection, client/contractor negotiations on the project effectively 'weed out' the less realistic offers of work. In the post-selection phase, selection is carried out on a more intensive basis to find the best contractor for the project as determined by the client/consultant criteria.

The nature of the bid and tendering system suggests that the process of supply in construction is the direct opposite of most of manufacturing industry. In manufacturing the producer determines the price of the good on the completion of production, so that it may be sold to the purchaser with a reasonable return or profit. In construction however, the client specifies the form of pricing, contract used and the final product (through the consultant) and the contractor puts a price to the client on this framework. An exception to this method of pricing is the management fee type contract, whereby a contractor agrees to carry out building works at cost, and in addition the contractor is paid a fee by the client. The contractor then provides management for the project and appoints sub contractors either on the basis of measurement or lump sum tender. However, even in this case risks may be high in comparison to other industries, since the management contractor assumes full responsibility for the control of the work.

The adoption of the tendering system of price determination may be attributable to two major points relating to the nature of the contractor's product. Firstly, within construction there is no standardised product that the contractor may sell (since in each bid the contractor sells differing aspects of service depending upon client demands), and therefore pricing must be done on a discrete basis for each product. Each time the contractor bids the firm is essentially putting a price upon a custom built final product that will only be of use to the client (or of limited use to others); the system thereby prevents wastage of resources and ensures a guaranteed demand for the final product so that the contractor can significantly reduce risk, and the client can get exactly what is demanded. A second point that has been instrumental in the adoption of the bidding process is the fact that the contractor can only sell the firm's services once in any project, and therefore must ensure demand for those services is acceptable to the firm before commitment to a specific project; once the service is sold it becomes embodied within the final product which is custom built for the client (Hillebrandt, 1985). Although it is possible for the client to resell the final product (probably at a lesser value to the purchaser than the owner of the final product for whom the product was specifically designed), the service within the product cannot be discerned from the final product, and therefore cannot be resold. This effectively means that the construction firm faces a temporary market for its services each time it is successful in a bid, which may be added as a cause of the acceptance of discrete price determination. It is argued here that this process of price determination may increase the level of risk in the industry beyond that in many others for the following reasons:

1. The contractor faces a different demand and supply schedule each time the firm enters a new contract, either in the pre-bid or bid environment. This leads the contractor to an indefinite situation over time such that the firm will have to employ a temporary workforce in many instances, perform to different demands and specifications, experience widely fluctuating costs and face an unknown future market (Park, 1970). This prevents the firm from pre-supply market testing of the contractor's product (ie services offered for construction) or establishing pricing policies and therefore the situation remains unpredictable for the contractor.

2. The contractor must include all his working and fixed costs into the bid price in order to maintain a return that may not be apparent or measurable to the contractor when the bid is placed. In addition to this, the contractor may be at a disadvantage to other contractors who can cover costs better or may be willing to accept a lower return. This makes demand for the contractor's services subject both to the relative and absolute cost of production, and therefore unpredictable over time.

3. The process of bidding will cost the contractor both in time spent and use of resources to prepare an accurate bid document, which will only be recuperable if the contractor wins the bid. Since this is by no means guaranteed, the contractor faces a potential loss situation each time he bids that may limit the firm's future ability to bid and work.

3.4 THE FINAL PRODUCT

The final product of construction activity which in terms of Figure 3.1 is the transformation of raw materials into a composite article via construction activity, is possibly the most influential factor upon the way the intermediate product of the contractor and structure of the industry have evolved. One of the major reasons for this is that the final product of the industry is the least flexible component of the characteristics involved in construction. There are two potential reasons for this:

1. The final product of the construction industry will be fixed to the point of utilisation since it will cover a large geographical area, be impractical to move for economic and technical reasons, and need to be produced with locational characteristics of the site of the product in mind. Because of this production will also have to take place at the point of utilisation.

2. A logical argument stemming from 1 is that the final product cannot be stockpiled as in other industries, and therefore the firm can only produce when a client demands a specified product at a specific location.

The immobility of the final product which prevents movement of the product to site will also ensure that the purchaser pre-demands the product in relation to supply, which effectively means that the client buys the final product before it is produced (Park, 1970). It is argued here that it is unfeasible for supply of the final product to take place before demand for the product in construction for the following reasons:

1. The final product if pre-supplied may not take into account client specific demands in terms of location and use of the final product. We have already suggested that the final product of construction is intermediate in that it is used for the creation of additional goods and services, and therefore is demanded as an investment good by the client. It is therefore the case that the client will only demand a final product that is of practical use to him in terms of location and product utilisation for the required task. Since these factors are likely to be diverse among prospective purchasers of construction products, it is unlikely that the contractor will successfully predict demand trends in all but very general construction (ie speculative construction).

2. Since the final product is 'large, heavy and expensive' (Hillebrandt, p.8 1985) a pre-supplied construction product that does not fit in with a prospective purchaser's demands and locational choice is unlikely to be adaptable to suit the client, on economic and technical efficiency grounds (for example, it would be unfeasible to move a building to the preferred site of a client). If we further argue that the contractor's working capital would already be tied up in the product anyway, it is unlikely that the contractor would use more capital for adaptation

without guarantee of definite demand for that product. Pre-demanded construction products dispel the need for this adaptation, and minimise uncertainty between client and contractor.

3. Related to 2, if we argue that the cost of the final product is high in relation to other goods because of the cost of land, materials and contractor's resources involved in construction of the product over time, it is unlikely that many contractors could afford to finance construction through working capital for more than one or two projects per year.

It can further be argued that from the client's point of view demand for a construction product will not be motivated by persuasive marketing by the contractor. It is unlikely that a prospective purchaser will be prepared to invest a large financial sum in a construction project at a specific location if the benefits of such a venture are not obvious to him. For this reason demand will come from the prospective purchaser according to the benefits of that investment at a specific location rather than any persuasive methods that a contractor could undertake. This may be added as further justification for pre-demand construction in the non-speculative case.

3.5 DEMAND AND SUPPLY IN THE CONSTRUCTION INDUSTRY

Given the nature and characteristics of the stages of construction as outlined in Figure 3.1 and above, we can analyse how these characteristics combine to influence demand, supply, and structure within the construction industry. If we regard the work of the contractor as a supplier of expertise (Hamman, 1971), there are two factors within the industry that may be taken as a function of the interaction of physical characteristics of the final product and the services offered by contractors. The first of these is that the production process, rather than the product of the industry, is the mobile factor in the relationship. While this has been attributed to greater mobility as a result of lower-capitalisation than in other industries (Neo, 1975), it should be noted that it is the highly mobile nature of production and the immobile product that are the determining factors in this situation. Barna (1983) has suggested that it is the mobile nature of the production process that provides the explanation for the low level of fixed capital assets within the industry (which may additionally be instrumental in the ease of access to the industry), while Hamman (1971) suggests that it is for this reason that mass production of a standardised product cannot take place.

By having a mobile production base, the contractor faces a higher level of risk than in other industries. Park (1970) puts this down to exogenous factors that cannot be catered for within the production process. These include for example weather and soil conditions that may adversely affect production by slowing up construction of the final product or even stopping it altogether. The fact that the final product is constructed outdoors in different locations therefore means that production becomes less predictable than in many other industries, and this may be added as an extra dimension of the final product that discerns the industry from others.

Since we have already argued that the final product cannot be stockpiled and the production process is mobile, it follows that mobility is the key issue in the success of the contractor to maintain production. However, in terms of demand within the industry, it is argued here that it is characteristics of both the product and production process that have had the major influence upon how demand is put to the contractor and industry. The fact that the final product is an investment good and is immobile prevents the pre-demanding of the final product, while the lack of a standardised product prevents the use of marketing techniques to create demand for the same product. It is the nature of the product which suggests that the purchaser will demand the final product according to his use for that product and location, and that these will be the major determinants of client demand.

In the context of this argument, it is valid to suggest that demand for the individual firm will be judged on two parameters:

1. Expertise of the contractor. Demand for the contractor will ultimately rest with the expertise that the contractor offers, since this is the product of the contractor required by the client to assemble the final product. For this reason demand is likely to revolve particularly around this aspect of construction. However, as we have already argued the term 'construction industry' covers many sub industries which may be distinguished by the type of expertise that firms in the industry offer as services to the client, due to

the fact that in construction the contractor offers the intermediate production process that may be classed as a service. Demand is therefore likely to be concentrated according to the needs of the client into a particular sub industry that forms the majority of construction work in that project.

2. Location of the contractor. The client will demand the final product at a given site at which the product is put to best use for that client. The client is therefore likely to favour a local construction firm for two reasons. Firstly, it will be cheaper to use the local firm than search for a contractor outside the market and bring in that contractor for the job, and secondly the local contractor will have more expertise of local exogenous factors (for example soil conditions) and may be able to account for these in the production process more than the 'non-local' contractor (although full accounting of these factors is impossible due to the unpredictable nature of exogenous factors in construction). Hence the client may favour a contractor already working in the location of the proposed final product on efficiency grounds.

It should be noted, however, that these two factors may not always interact, and that the second condition presupposes that the relevant demanded skill exists in that location or market. If the client requires a final product that demands a highly specialised skill for construction (eg a petrochemical plant) then it is unlikely in many locations that the local contractor will have the expertise to carry out the construction of the product. In this case expertise will become more important than cost considerations and a contractor will be sought on the criterion of expertise rather than knowledge of local conditions. While it should be noted that this is possible in contracting because of the mobile nature of the production process, it also illustrates an important factor of demand for the individual firm. The more specialised the skill required for construction of the final product, the less likely are factors other than expertise to influence demand for that firm's product. This occurs because the client ultimately requires the final product as an investment good given the location of the product, and therefore the contractor will be demanded on expertise grounds specifically. Since demand for specialist construction is likely to be limited in any one market, a corollary of this is that the more specialised the firm, the more mobile it must be between markets to maintain

An Economic Overview of the Construction Industry

demand for the skills it possesses. The mobility of the contractor in this case is also an advantage if we argue that the nature of the product suggests that each contract will often represent a large amount of the total work of the contractor in any year, and therefore there will be considerable discontinuities in production functions as a consequence of this. Mobility of the contractor may lessen this problem by enhancing production prospects in all regions and therefore decreasing the unpredictability of demand.

In terms of industry supply, the analysis is similar to the determinants of demand as discussed above. Hillebrandt (1985) suggests that supply of contractors for any available work can be put down to two basic characteristics:

1. Size of contract. A large contract will clearly not only require a greater input of resources of the contractor than a small contract (in terms of organisation and management as well as technical skills and plant), but will invariably need a more detailed and costly bid preparation if the firm is to compete feasibly at the tender stage. Since financial and technical capability may be positively related to size of the firm, the number of contractors competing for any project will be restrained by both size of contract and firm in any situation. It should however be noted that this argument will work two ways; the larger firms will have greater overhead costs than smaller firms as a consequence of having larger production and managerial resources and therefore can only compete on projects that offer the chance of covering overheads and making a return. This is only likely where the contract size is of sufficient size to cover these factors, so that large firms will generally not tender for small scale projects on economic efficiency grounds.

2. Complexity of contract. The more complex the contract, the fewer contractors are likely to supply the demanded services for two reasons. Firstly, the more complex contract will require greater management skill and expertise for coordination which will require detailed knowledge of the service demanded, and secondly the expertise and know-how related to construction of the final product is likely to be available only within a small number of firms because expertise is related to past knowledge and training that will not be easily accessible within the industry. Both factors are likely to limit the supply of firms in this situation, so that the more complex the contract the

more likely it is that firms in a specialist sub-market will be involved rather than general contractors.

If complexity and size are combined as the determinants of supply of services then clearly supply in the whole industry is highly fragmented and diversely split into sub-markets so that the number of firms interested in any one contract may be small relative to the number of contractors in the industry. We can further argue that the larger and more complicated the contract, the less influence will location of the contractor have (as suggested by demand conditions) and the lower will be the number of interested firms in any sub industry, which will also limit the supply of contractors. These may be added as further dimensions of fragmentation that occur because of expertise and spatial considerations. The similarity of this situation of supply with the demand environment in the industry illustrates the close interlinking of the concepts that may be attributable to the specific nature of the product and production process of construction activity.

3.6 STRUCTURE OF THE CONSTRUCTION INDUSTRY

The structure of the industry may be regarded as a reflection of the demand and supply situation in construction that occur because of the nature of the product. The relationship between industry structure and demand and supply determinants will however be complex. The direction of causality of influence between these factors may be unidentifiable because of the interrelationship that exists in a dynamic context between industrial structure and the demand and supply environment.

One clear factor concerning the structure of the construction industry to have emerged from the discussion of demand and supply above is that the industry as a whole covers many sub industries that are determined by expertise, size of the firm, and location (though this third influence may be questioned in relation to the other two in certain circumstances). The size of the market will relate both to demand for the final product of that sub industry and supply of the contractor's services which may be influenced by geographical or specialist knowledge limitations.

We argue here, however, that there are several other factors that must be taken into account in the discussion of

An Economic Overview of the Construction Industry

the structure of the industry that may affect the demand and supply situations. These are:

1. barriers to entry and access to the industry;
2. client influence and differentiability of the product;
3. level of competition within the industry.

These are now discussed in turn.

3.6.1 Barriers to entry and access to the industry

From the discussion in section 3.3 above we can argue that the nature of the construction industry is such that freedom of entry is not restricted to any significant degree, which is attributed to the nature of the contractor's product. Barriers to entry within the industry are not likely to prevail to a significant degree for five reasons:

1. Costs of entering the industry will not be high relative to other industries because the contractor relies upon human capital (as argued in section 3.3), and therefore at the lower end of the market the contractor may enter the industry armed only with expertise and know-how. Since growth of the firm involves the accumulation of expertise and know-how, this feature of the industry prevents the erection of barriers to entry since the knowledge can easily be transferred.

2. In connection with 1, because the product of the firm is expertise and know-how, any barriers that existed could be vaulted by a potential competitor by buying expertise in the form of management and technical staff. The level at which this firm enters the industry clearly depends upon the financial resources available for the procurement of the expertise.

3. Because price determination is carried out on a discrete project to project basis, the opportunities for limit pricing strategies within the industry are constrained. Although firms could feasibly collude to prevent one firm from winning a bid by tendering unprofitable prices, the lack of additional barriers to entry to prevent other firms entering the market would ensure that this barrier would be impossible to maintain as a significant deterrent to entry.

4. Advertising will not provide an effective barrier to entry because the product is pre-demanded and therefore the contractor will tend to advertise abilities of the firm

rather than promote itself on a project to project basis. The entrant can therefore feasibly compete with the established firm on a bid on the grounds of technical expertise.

5. Economies of scale of established contractors will be unpredictable because the final product characteristics change with each project, so that this will not provide a major barrier.

Hillebrandt (1985) argues that the effects of having insignificant barriers to entry are twofold: firstly, the small firm that is efficient may graduate up the scale fairly easily without the threat of anti-competitive practices, and secondly that the ease of entry into specialist markets has prompted the larger firms to move sideways into various other specialist markets as a means of stabilising demand for the firm's services, and taking advantage of high demand markets. It should be noted that this second argument may also be important in locational choice - the contractor will be free to move from one location to another in the absence of effective barriers to entry because of mobility of the contractor's product and lack of impediments that occurs through lack of barriers, though this clearly depends upon demand at that location; it is a consequence of client led demand that the contractor will only locate where there is demand for construction services because the firm cannot generate demand to any significant degree at a given location and therefore must follow market trends. We therefore argue here that barriers to entry within the construction industry are not effective enough to prevent entry of new firms to the industry, and that this is a consequence of the contractor's product and the final product of construction activity; we further suggest that the lack of barriers is instrumental in the contractor's choice of markets in terms of location and type of product offered, and that this has obvious effects upon demand and supply in the industry.

3.6.2 Client influence and product differentiation

The nature of demand and supply within construction is not something that easily fits into economic theory; the final product is pre-demanded and sold before production begins, and supply for that product is to designs and specifications given to the contractor by the client (cf Figure 3.2). Given this situation, and that as we have already argued the contractor provides a service, it is reasonable to argue that the contractor's product will be very similar to that of competitors (ie construction of a given final product). Within construction, therefore, it is in the interests of the contractor to aim to differentiate his product to that of competitors as a means of drawing attention to the contractor's offer in the pre-bid state. The final way in which the contractor differentiates himself from others is of course price, which provides the basic criterion for contractor selection in the bid and negotiation process, but other forms may also take place; expertise may be considered a form of differentiation, but will only serve to differentiate sub industries rather than individual firms unless that expertise is firm specific only. The need for firm specific differentiation has led to the adoption by contractors of various marketing practices in which the traditional role of the construction firm has been superseded. We have already mentioned the introduction of management fee contracting that may be considered a means of product differentiation, but we include three further types of construction activity:

1. Design and build; the contractor carries out both the design of the individual project and the on-site construction work. In that the design and construction process will be more integrated than in separate design and construction, this usually results in lower production costs on site, a shorter contract period and overall price saving to the client although requested variations by the client are usually discouraged to protect the contractor's implied warrant of suitability.

2. Package deal; in the package deal the client chooses a suitable building almost from a catalogue of highly standardised products that the client designs and builds, usually from experience. Since the building is ultimately 'off the peg' construction is very quick, but there is little room for variations should the client dislike certain aspects of the design.

3. Turnkey; this form of construction requires the planning, design, construction, furnishing and equipment to be undertaken by one firm. An agreement is thus made with the contractor to provide an all embracing contract. This type of contract is usually adopted for highly specialised types of project and will be custom built for the client, although client control of costs quality and structure are somewhat limited.

These examples illustrate the potential for product differentiation within the industry. We have already argued that product differentiation may not provide a significant barrier to entry into the industry, but it is suggested here that this process may be instrumental in persuading the client to put the contractor on the tender list, and get the contractor noticed within the bid. For this reason we argue that the contractor will aim to differentiate the services the firm offers from others in an attempt to motivate interest in the contractor, and that this will be a consequence of client led demand.

It should be noted within this discussion that the actions of the client must also be taken into account in the process of competition. If it is argued that the client is a monopsonist, since almost all construction projects are let on a separate basis, then the client can influence the competitive situation. In that demand is client led, the firm is given a 'take it or leave it' ultimatum to bidding. This enables the client to not only choose the method of pricing and specification of contract, but introduces the possibility of the client demanding more than the construction of the final product; the advent of turnkey projects for example may be taken as either a move by the contractor to differentiate his product, or demand initiated by the client in response to monopsonistic power. From this standpoint we can suggest that while product differentiation is likely in the construction industry, the initial incentive for this, which may affect structure of the industry, is indeterminate since it is potentially a response to the interaction of the demand and supply sides of construction activity.

3.6.3 **Level of competition**

Given that there are no effective barriers to entry and product differentiation may be justified, we now turn to analyse the structure of the industry as given by level of competition. Table 3.2 illustrates the different types of competition that may ensue given the method of price determination and the prevailing market conditions in that sub-industry or industry. Where there are many firms in the market, competition may range from that approaching perfect competition to monopolistic competition, although the client may prevent this from occurring by adopting a different form of price determination that does no

An Economic Overview of the Construction Industry

emphasise the need for monopolistic competition. In the lower half of the table we see that where there are few firms in the market an oligopolistic type situation generally prevails, because the firm must take account of other contractor's actions. However, this may not be deemed a true oligopoly since none of the firms have a large enough share of the market to make a significant contribution to total market output, oligopolistic influence is limited by client power and barriers to entry may not be efficient.

The table illustrates that the level of competition will vary according to the nature of price determination rather than barriers to entry, which is different to many other industries. In that price determination may take several forms which will be influenced by client led demand, the construction industry is differentiable from many others in structure as a consequence of demand and supply factors. If we further consider the physical nature of the final product, we can argue that the industry is unique in terms of characteristics because of the influence of the final product on the production process and industry structure. It therefore seems justified to analyse the industry within the framework of these characteristics on the grounds that the industry is differentiable from all others according to these factors.

3.7 ASPECTS OF INTERNATIONAL CONSTRUCTION

Although international construction can clearly be investigated within the context of the framework given above, there will be certain consequences of operating abroad that will not be relevant in the domestic contracting situation. It is argued here that these characteristics of international contracting may affect demand, supply, and the structure of the industry so that it may differ in certain respects from the outline discussed above. We distinguish three factors here that may be associated with international contracting:

1. Level of risk. Possibly the most obvious distinction between international and domestic contracting is the higher level of risk the contractor faces in the international market. While the introduction of unknown climatic and environmental conditions will clearly influence construction of a specific project in this sense, other factors associated with risk may also be relevant. One aspect of the risk faced

Table 3.2: Assessment of Type of Market in Contracting

Type of Selection	Stage of Selection	Number of Firms	Product Differentiation	Type of Market
i. MANY FIRMS IN THE MARKET				
Open tendering	Tender	Many	None	Approaching perfect competition
Selective tendering	Pre-tender Tender	Many Few	Substantial None	Monopolistic competition Partial oligopoly without product differentiation
Two-stage tendering	Pre-tender Tender	Many Few	Substantial None	Monopolistic competition Partial oligopoly without product differentiation
Negotiation	Negotiation Pre-selection Post-selection	One Many One	n.a. Substantial n.a.	Limited monopoly Monopolistic competition Limited monopoly

Table 3.2: continued

ii. FEW FIRMS IN THE MARKET

Open tendering	Tender	Few	None	Oligopoly without product differentiation
Selective tendering	Pre-tender	Few	Substantial	Oligopoly with product differentiation
	Tender	Few	None	Oligopoly without product differentiation
Two-stage tendering	Pre-tender	Few	Substantial	Oligopoly with product differentiation
	Tender	Few	None	Oligopoly without product differentiation
Negotiation	Negotiation	One	n.a.	Limited monopoly
	Pre-selection	Few	Substantial	Oligopoly with product differentiation
	Post-selection	One	n.a.	Limited monopoly

Source: Hillebrandt (1985) p.147

by international contractors is exposure to political risk that may include expropriation of the contractor's assets or nationalisation of the firm. Since contractors inevitably have to follow demand, it is likely that they will tend to locate mainly in the high risk LDC regions where demand for large scale construction projects is greatest and effective indigenous competition is lowest. By entering and setting up operations in these regions, the contractor faces the possibility of hostile customs and tariff barriers on plant and materials, xenophobic reaction to the firm, and hostile or prohibitive host government legislation that may ultimately lead to expropriation. While this may not present as great a threat to the contractor as a manufacturer because of the mobility of the contractor's production process, in the event of expropriation or nationalisation the contractor is likely to lose all plant and equipment in that country, plus any chance of return on sunk costs used in setting up or bidding, that may present a significant monetary loss.

In addition to this, the contractor faces risk as a direct consequence of undertaking international operations in any country. This will be due to differences in national financial systems and exchange rates that may prevent the contractor from obtaining a reasonable rate, or in some cases any return at all; exchange rate fluctuations, host country restrictive repatriation of profit policies, delayed or postponed payments and differential taxation and legal frameworks between countries all provide significant complications to international contracting that are not relevant within the domestic environment.

2. Project size and scale of operations. If we argue that the international contractor faces a higher level of risk than in domestic contracting, then clearly there must be some form of incentive to overseas work. It is suggested here that part of the motivation behind this is the larger scale of operations with which the firm will be involved in the international industry. In terms of demand for construction activity, in LDCs where demand is greatest scale of project is likely to be larger than in many DCs because much of the basic infrastructure is still required, and therefore scope for construction work is high, particularly if we consider that LDCs will generally not have an indigenous construction sector able to undertake such construction. Supply of contractors will take place according to expected returns. Overseas involvement will

not only have higher overheads and costs to the contractor as a consequence of entering the host market, establishing a presence and bidding for projects, but will also require a greater motivation in terms of higher returns to the contractor to offset the additional risks of such a venture. Since we can argue that these costs and returns are only likely to be covered on large scale projects, the international contractor will generally bid only for major works where a significant return can be made. Scale of operation is therefore likely to be higher in international contracting than domestic activity, as are costs and returns of the contractor.

3. Level of demand. From 2 above we can argue that demand in developing region markets is likely to be greater and more sustainable than any national market for construction activity, particularly if the international contractor is prepared to move around the various markets in search of work. While this may be seen as an advantage to the international contractor three factors should be taken into consideration:

   a. The higher level of demand is matched (if not superseded) by a greater supply of contractors than in the domestic environment, who will come from various countries and be motivated by demand levels and prospects of high earnings. In terms of level of competition therefore the situation may be worse than in many domestic situations.
   b. Type of demand and client will be more variable than in domestic construction as a consequence of different locations and environments. When coupled with the greater number of competitors in the industry, this may make the contractor's chance of success at bidding slimmer than in the domestic situation. If we further argue that bidding cost will be positively related to cost (Neo, 1975) and scale of construction is likely to be large (from above), significant losses from bidding may be added as a further aspect of risk in international construction.
   c. The high number of contractors within the industry has led many clients to adopt policies of restraining competition to manageable levels in tendering. These policies include prequalification (whereby the consultant chooses the most able contractors to bid based upon reputation and

experience of the contractor to eliminate wasteful tendering) and the provision of performance and guarantee bonds (where contractors are demanded to give the client monetary bonds that may be called against the contractor if that aspect of the construction stage is unacceptable to the client). The most common forms of bonds are bid bonds (to ensure that bids are serious and the contractor will undertake the work if required), and performance bonds to ensure quality of the supplied construction project.

The three factors given above which distinguish international contracting from domestic contracting are not unique to the industry, but may be common to international business (although there are clearly industry specific influences on the environment). In this respect, the difference between national and international contracting is similar to any other industry. However, in that the domestic situation is different from other industries because of the characteristics of demand, supply, and industrial structure (that reflect the nature of the product) clearly the international contracting industry will be different from all others in this respect also. If we therefore combine the international nature and problems of international contracting with industry specific characteristics (that are also reflected in domestic construction) we get an industry that is unlike any other, including domestic construction; this provides justification for the research, and suggests that there will be industry specific traits associated with the international involvement of contracting firms.

Chapter Four

APPLICATION OF MNE THEORY TO INTERNATIONAL CONSTRUCTION

4.1 THE RESEARCH FRAMEWORK

For the purposes of this research, the eclectic paradigm as set forward by Dunning (1981, 1985) provides the theoretical framework upon which the analysis of the international contracting industry is based. The choice of this framework reflects both the inability of the various strands of MNE theory to encompass all aspects of multinational involvement, as suggested in Chapter 2, and the need for a general rather than specific theory of the MNE to incorporate aspects of international contracting that are unique to the industry into the research (as specified in Chapter 3). The eclectic paradigm suggests that there are three necessary conditions for FDI to take place:

1. The firm possesses competitive or ownership advantages over firms indigenous to the host country, and also over firms of other nationalities.
2. Ownership factors are more advantageously exploited internally by the firm rather than externalised by means of selling or licensing these advantages to other firms.
3. Assuming conditions 1 and 2 are satisfied, it is more profitable for the firm to undertake production outside national borders than it is to service foreign markets by domestic production and export.

The greatest advantage the eclectic approach has over others, particularly in the empirical study of MNE activity, is its generality; by suggesting that international production takes place because of the presence of ownership (O), locational (L), and internalisation (I) factors, the approach provides justification for the existence of MNEs according to both endogenous and exogenous factors that may be reflected within the conditions of the framework.

The major factor about the OLI paradigm is that if one or more of the conditions are missing, the firm will not undertake FDI, or even search for overseas markets in some cases (ie where locational factors do not warrant it). By specifying the need for simultaneous attainment of the three conditions the framework does not automatically assume that these factors are fixed, but enables them to vary with firm and industry specific characteristics. This has been useful in the past for the analysis of industries other than MNE manufacturing (upon which most MNE literature is based); Dunning and Norman (1979) for example use the approach to explain the determinants of multinational companies office locations, while Dunning and McQueen (1981) apply the eclectic theory to the international hotel industry. The notion that FDI may be explained by three apparently simple conditions is appealing in terms of empirical study - the logic behind the conditions (that may be attributed to economists working in the three related fields as well as Dunning's approach) as a prerequisite for FDI provides a basic though powerful framework upon which several aspects of the MNE may be analysed within the OLI parameters. Hence the eclectic approach may contribute, say, to the understanding of Vernon's PC model by introducing internalisation and the interaction of ownership and locational factors into the analysis. As such, the eclectic approach does not suggest that there are totally new aspects of MNE activity to be considered, but that the existing literature be redefined as interrelated parts of the overall picture. In this sense, Dunning does not reject any of the existing theories of the MNE within the OLI paradigm, but places them within a framework whereby they may be justified according to factors relating to firm, industry, and home and host country characteristics that are encompassed within the OLI conditions. It is for this reason that the approach provides the most flexible basis for empirical study of the MNE, and as such is the theoretical foundation upon which this research into the international construction industry is based. The rest of this chapter therefore sets forward the theory in more detail and illustrates how this fits in with characteristics of the industry as outlined in Chapter 3.

4.2 OWNERSHIP ADVANTAGES

The initial principle behind O advantages is relatively straightforward and provides the basis for the industrial organisation approach to FDI. In order for a firm to compete with foreign enterprises in their own market as an MNE, the firm must possess certain advantages to compensate for additional costs of selling or producing in an alien environment.

Dunning (1977) specifies three types of advantages that may accrue to an enterprise of one nationality over that of another:

1. those which may arise as a consequence of ownership of proprietary assets, regardless of multinationality;
2. those which a branch plant of an established national enterprise may have over de novo firms;
3. those which occur specifically because of multinationality.

Since the ownership advantage approach in this context is similar to the industrial organisation approach to the MNE, it is to be expected that several of the advantages that Dunning specifies relate to earlier works in this field. Dunning's advantages generally reflect the work of Hymer (1976) and variants upon this work put forward by Kindleberger (1969), Hirsch (1976), Caves (1971), Aliber (1970), and Johnson (1970).

The notion of O advantages, however, has subsequently been modified (Dunning, 1985) to distinguish between asset advantages and transactional advantages of MNEs that are a function of structural market imperfections and transactional market imperfections respectively. While structural market imperfections particularly relate to the industrial organisation approach as outlined above, transactional advantages relate more to the benefits (or lesser costs) of the internal organisation as compared to the external market, and therefore clearly corresponds with the benefits accruing to internalisation. Dunning distinguishes three types of transactional advantage:

1. those that reduce risk and uncertainty that may be present within international operations (as identified by Vernon, 1983);
2. those that maximise plant economies of scale in

imperfect product markets;
3. those that capture benefits (or reduce costs) of externalities and imperfect individual transactions.

By specifying O advantages into these groupings, Dunning is implicitly acknowledging that ultimately there will be three factors relevant to the generation of O factors, firm specific, industry specific and country specific characteristics. If these factors are assumed to be the major determinants of O advantages, the propensity of the firm to engage in international production depends both upon industry specific factors and on the ability of the firm to generate O advantages that are best exploited within the internal organisation in foreign markets. Since this will ultimately come down to industry specific factors affecting both firm specific and country specific advantages, we now turn to the relationship between these components of O advantages in terms of international construction.

4.2.1 Firm Specific Advantages

A firm specific advantage may be regarded as an O advantage that is exploitable by an individual company only, and therefore not available for use by others unless directly specified. As such, the possessing firm may have definable property rights over the advantage that prevents others from usage, although generally other firms in the industry may have similar advantages, depending on the nature and source of that advantage. Firm specific advantages often cited include brand names, size of the firm and R&D capacity, that the firm may use to compete in foreign markets with both indigenous and other foreign competitors.

In terms of firm specific advantages within international contracting, two factors are worth noting:
1. Unlike multinational manufacturing, international construction involves a service that is essentially not capital intensive. As such, R&D expenditure will not provide a significant advantage as in capital intensive industries, but O advantages will revolve more around human capital aspects.
2. Within the industry the need for firm specific advantages arises from competition with other foreign contractors in overseas markets rather than indigenous contractors. This may be put down to the lack of experience

Application of MNE Theory to International Construction

and expertise on the part of indigenous contractors in the major overseas markets (ie LDC regions).

The basic motivation for the contractor to generate firm specific advantages will be to differentiate the firm from others in the market. This may be a consequence of the large number of contractors in the industry, but may equally be attributed to the mechanism of obtaining work in international contracting (ie the bid system); it has been pointed out in the preceding chapter that the bidding system provides the most efficient way of matching client demands with contractor supply given the nature of the final product. While this system tends to suggest that competition will be based solely upon price, it also paradoxically implies that the contractor must compete on more than price to be recognised within a bid. By differentiating the services offered from others, the contractor's bid is likely to stand out as more attractive to the client ceteris paribus, and therefore stand a better chance of success. As such the ability of the contractor to differentiate the firm from others provides a major firm specific advantage that effectively permits that contractor to compete against international contractors in overseas markets. This may be carried out in two ways. The first strategy the contractor may adopt is differentiation of the firm itself. Because the product of the contractor is embodied within the final construction project, the contractor may utilise reputation of the firm based upon past experience as a major source of differentiation. This is necessary in this case because the contractor's product will be embodied within the final product, and therefore cannot physically be illustrated to the potential client. Reputation of the contractor will embody many aspects of the contractor's activities, especially past experience of the firm, specialist expertise of the contractor and quality of past work. This in turn will be reflected in the contractor's name, that may be regarded as the brand name of the contractor in the bidding situation. The contractor whose name carries a good reputation is likely to be more successful than a lesser known contractor for several reasons:

a. The contracting firm, by using its name, automatically guarantees a specified quality of construction to the client. If this standard is not attained the contractor is likely to suffer a loss of reputation that may impair future bidding. For this reason the client may be assured of an acceptable minimum standard of work that is not necessarily the case if the contractor is not well known.

b. Reputation not only allows the client to assess the contractor's work first hand (ie past works completed), but also suggests technical expertise of the contractor via experience. On this basis the client is likely to prefer an experienced contractor as a further guarantee of expertise.

c. In many international tenders prequalification is imposed by the client's consultant to prevent unrealistic bidding. A major criterion for success in prequalification is past expertise and reputation (as encompassed within the contractor's name) and therefore a well known contractor will benefit in prequalification over a lesser known contractor. A similar case may be made for selective tendering.

The second means of differentiation is through the product itself. Given the nature of the bidding system, clearly the most obvious way that the firm can differentiate its product is via price. In general the cheaper the relative price of the contractor's tender in the bid situation, the more will that tender be favoured. However, there are other sources of product differentiation that the contractor may exploit. One way which has been outlined in Chapter 3 is the ability of the contractor to offer specialist construction skills rather than maintain workload on basic construction work only. This may confer two benefits on the contractor:

a. If we argue, as suggested in Chapter 3, that the more complex the construction demanded for a project the lower the number of contractors technically able to carry out the work to the required standard, then clearly by entering a construction sub-industry the contractor will have a greater probability of success than in a less sophisticated market.

b. The contractor is likely to benefit in all markets from diversification of specialist skills, because diversification in this sense will lead to an increased and broadened financial base, enhanced reputation across the market generally, and the ability to carry out projects requiring specialist skills without having to sub contract. In the context of Dunning's O advantages, this may potentially be regarded as a transactional rather than structural advantage, in that the contractor benefits as a result of internal diversification.

The relative ease with which contractors may enter specialist sub industries can be put down to the mobile nature of the contractor's product; the emphasis upon human rather than technical capital means that a contractor with adequate financial resources can 'buy in' expertise and automatically enter that specialist market and compete. This may be considered a consequence of international construction being a service industry, and suggests that artificial barriers to entry of sub-industries will not present a major problem to contractors.

A second way the contractor may differentiate a firm's product is by offering additional factors over and above that of construction of the final product which generally will relate to services that normally take place before or after construction activity. Hence the contractor for example could offer finance for the project, maintenance of the final product or training of personnel to use the final product (in the case of industrial construction, say) as a source of product differentiation. In this sense, the contractor is undertaking a form of vertical integration and diversification. One way in which this may take place is by the contractor moving backwards into design in addition to construction. It has already been suggested in Chapter 3 that design and construction are often two separate parts of the same process, and therefore the contractor may aim to get more involved in both phases of production as a means of minimising client transactions with separate bodies in the industry (and therefore reduce complications for the client making the package more attractive). The advent of design-construct packages and turnkey operations in particular may be considered part of this, and regarded as an aspect of the firm undertaking specialist expertise to differentiate its product. In this respect, vertical diversification and the entry of specialist construction market are similar, which suggests that these may be used simultaneously to enhance the contractor's product overall.

It should be noted that differentiation of the firm and the firm's product are likely to take place alongside each other, and may not be easily discernible in practice because of the interaction of the two; expertise for example may be regarded as a source of product differentiation because it implies a superior product, that will win the firm more work and enhance the contractor's reputation. As such it is both a product and firm differentiable factor.

There are potentially certain requirements for both types of differentiation that we predict will form the major firm specific advantages in international construction because they will enhance and increase the ability for

differentiation of the contractor overall:
1. Name of the firm. The firm's name embodies and represents reputation, past experience and specialist expertise of the contractor. As such, it will be a major source of firm specific differentiation because it enables the contractor to effectively compete against all others in the industry by differentiation of the firm in the bid situation. As the name of the firm will encompass all relevant aspects of the contractor's competitive performance, this may be classed as the major firm specific advantage in terms of differentiation of the firm in the bidding situation.
2. Human capital. Contractors are likely to place a great deal of emphasis upon expertise of their workforce because this will be the source of the firm's name and reputation. Firm specific training of personnel for overseas work is therefore likely to provide a significant firm specific advantage in this respect.
3. Services offered by the contractor. In terms of both technical knowledge and vertical diversification, international contractors are likely to aim at diversifying their services for the reasons outlined above.
4. Size of the firm. Generally the larger the enterprise, the more access will that contractor have to cheap finance (via either the loan market or internal funding), better production resources and the like. This will not only enable the contractor to bid for larger contracts (as suggested in Chapter 3), but will also give the contractor the means to acquire a competent workforce, and diversify into technical and construction related services, as mentioned as firm specific advantages in 2 and 3 above. As this may also enhance the name of the firm, size of the contracting company is likely to provide a significant firm specific advantage.

4.2.2 Country specific factors

Country specific advantages may be defined as advantages that occur because of the nationality of the enterprise, and therefore will be exploitable only by firms of that particular nationality. The basis for these advantages will be influences of the home country of the firm (or the interaction of home country and host country factors) on O advantages that are not freely transferable to other nationality firms. The implication of this in the Dunning model is that firm specific advantages will reflect country specific factors, although the degree of this depends upon the advantage in question.

Table 4.1 gives an outline of the most commonly suggested O advantages and the country specific factors which may generate and sustain them as given by Dunning (1981). This table shows for example that it is postulated that size of the firm (which we may regard as a firm specific advantage as outlined above) is likely to be positively related to structure and size of the markets of the firm and lack of legal restrictions on growth, while the ability of the firm to undertake product differentiation is determined by willingness of the market to accept and sustain persuasive marketing methods. If, as Table 4.1 suggests, country specific endowments are relevant in the creation and maintenance of O advantages within an industry, and success in international markets is attributable to the successful exploitation of O advantages in foreign locations, then it is logical to argue that relative success of the individual firm will ultimately come down to nationality of that enterprise if MNEs predominate in that industry. This will clearly be of particular relevance to any industry specific analysis within the context of this theoretical framework.

One country specific factor that may be highly relevant in the generation and exploitation of O advantages in most industries is home government policy which affects the competitive stance of firms in that industry when matched against other nationality competitors. On the question of government influence on O advantages within the eclectic theory, Dunning suggests that: 'Government intervention ... affects both the generation of ownership advantages and the economic ties between investing and recipient countries.' (p.89, 1981).

The implication of this statement is that there are two ways in which the home government may affect the MNE abroad. The first is by legislation and action that directly affects the nature of O advantages. This may include government participation in the industry, availability of skilled workforce, ability to differentiate products, regulations concerning size of the firm, and other possible interventions within that industry. The second influence relates to indirect effects that will influence the success of the firm in particular markets. These will depend partially

Table 4.1: Links Between Selected Ownership-Specific Advantages and Country-Specific Characteristics Likely to Generate and Sustain Them.

Ownership-Specific Advantages	Country Characteristics Favouring Such Advantages
1. Size of firm (e.g. economies of scale, product diversification)	Large and standardised markets. Liberal attitude towards mergers, conglomerates, industrial concentration.
2. Management and organisational expertise	Availability of managerial manpower; educational and training facilities, (e.g. business schools). Size of markets, etc, making for (1) above. Good R&D facilities.
3. Technological based advantages	Government support of innovation. Availability of skilled manpower and in some cases of local materials.
4. Labour and/or mature, small scale intensive technologies	Plentiful labour supplies; good technicians. Expertise of small firm/consultancy operation.

5.	Product differentiation, marketing economies	National markets with reasonably high incomes; high income elasticity of demand. Acceptance of advertising and other persuasive marketing methods. Consumer tastes and culture.
6.	Access to (domestic) markets	Large markets. No government control on imports. Liberal attitude to exclusive dealing.
7.	Access to, or knowledge about natural resources	Local availability of resources encourages export of that knowledge and/or processing activities. Need for raw materials not available locally for domestic industry. Accumulated experience of expertise required for resource exploitation/processing.
8.	Capital availability and financial expertise	Good and reliable capital markets and professional advice.
9.	As it affects various advantages above	Role of government intervention and relationship with enterprises. Incentives to create advantages.

Source: Dunning (1981) p.87

upon the ability of the individual firm to enter markets, but also upon the relationship between the home and host country of the MNE in terms of cultural, economic, and political ties that may be attributed to government actions and influence. The implication of this is that success in a specific international industry may be related to nationality according to home government policy of those competitors, and additionally that there may be patterns in regional markets whereby firms of one nationality are more successful than firms from other countries because of home and host country relationships, particularly in terms of government influence.

The exploitation of country specific factors within international contracting is likely to follow similar justification as that given above concerning firm specific advantages in the industry: it is in the contractor's best interests to differentiate the services the firm offers from others to 'stand out' in the bidding situation. Although firm specific advantages will represent the individual O aspects of the firm, it is likely that these will reflect factors associated with the contractor's nationality so that firm specific advantages will generally be similar (though obviously not exactly the same) for contractors originating from the same country. Country specific advantages may therefore be of great importance both to the contractor competing with other nationality firms, and the national group of contractors in the international market. The basis of competition on nationality is clearly something that is avoided in the domestic market, and therefore may be added as a major difference between the domestic and international construction scene.

In consideration of the fact that firm specific advantages will generally reflect the home country environment to some extent, three country specific characteristics that are likely to have a major effect on the generation of O advantages in international contracting are given below:

1. Size and nature of the domestic market. Size of the domestic market will be beneficial to contractors from that country for two major reasons. Firstly, size of the domestic market is likely to be positively related to average size of the construction firm, so that the largest national market will contain the largest contractors (which as already suggested forms a major firm specific advantage). Secondly, the larger market will have greater prospects of

demand for construction works, and therefore will give the contractor the opportunity of training and acquiring expertise and experience that will be necessary in the international environment. Hence contractors in large domestic markets will be able to exploit size and expertise more than contractors from smaller countries, although this advantage may be overstated.

Size of domestic market will also be related to nature of demand in that market (and consequently level of development of the indigenous construction sector), that will be a function of relative endowment of resources in that country. Dunning (1981, 1985) suggests that the relative resource endowments of a country will be reflected in the state of the domestic market, so that a country well endowed with human capital is likely to acquire O advantages in knowledge intensive sectors, while countries endowed with a large unskilled or semi skilled sector will develop more labour intensive advantages. In terms of construction activity, this suggests that the nature of the final product demanded will determine the contractor's skills required, so that a capital intensive industrially developed country is likely to require high technology industrial construction and a labour intensive low technology country will require labour intensive construction. In that domestic firms will adapt their expertise according to domestic demand (assuming the indigenous construction sector exists), the relative resource endowments may therefore reflect firm specific advantages through country specific factors. Since these factors will influence both level of expertise and diversity of skills in the domestic market (that will be positively related to size of market), relative resource endowment may provide a significant factor in the development of a country's construction skills that generate O advantages for overseas work.

If the domestic market is both large, and by nature specialised in areas that are likely to be highly demanded in overseas regions, then this situation may confer a transactional advantage on all contractors of that nationality in the overseas market; the more that nationality's contractors are demanded and enter the international market, the more the reputation of all contractors from that country will be enhanced to favour the whole group within the industry. This positive externality may be added as an additional advantage of size and nature of the domestic market.

2. Demand for related services. In connection with 1 above, the more successful abroad are home country companies whose product is related to construction, the more likely are contractors of that nationality to benefit from this in those markets. The most obvious example of this is demand for consultants on construction projects. A contractor of the same nationality as a consultant appointed to select favourable bids may stand a better chance of success in that situation, not necessarily for nationalistic reasons on the part of the consultant, but because adopted procedures and specifications in the bid will be efficiently catered for by the contractor on the basis of experience.

Demand for home country products other than construction may also be a significant factor in terms of generation of O advantages; generally the more home country enterprises that move abroad as a response to foreign demand, the greater will be the demand for home country contractors by these enterprises to build production facilities where necessary. Home country firms are liable to require home country contractors because this will minimise costs of searching for a contractor, and will guarantee a certain level of construction on the basis of domestic knowledge of the contractor (either at first or second hand).

Demand for home country products related to construction may be considered an advantage in that they stimulate demand for home country contractors, and therefore as argued in 1 above, provide a positive externality by means of enhancing reputation. In addition to this, this may enable contractors to enter new regions and automatically compete with established (indigenous and foreign) contractors.

3. Home government support. Government support, where available, is likely to have a major influence on success of contractors abroad in both the direct and indirect sense as discussed above. Direct government support may take the form of specific support and help for contractors via financial or technical assistance, the promotion of contractors' interests abroad, or the development of a strong national industry with which to compete abroad. Clearly the level of direct government support depends upon the political stance of that government, and the resources made available for direct help. As such, direct government support is likely to vary between countries.

Indirect benefits will come from general factors such as the relationship of the home and host country in political terms, past relationship of the governments and the like. In addition to this historical relationships between the countries may provide a significant country specific advantage. These types of benefits will vary according to the home country and the host market in which the contractor enters, and also according to the relevant political and economic environments in the countries over time.

4.2.3 Oligopolistic reaction within international contracting

The role of oligopolistic reaction and strategy as motivation for retaliatory FDI along the lines of Knickerbocker (1973) and Graham (1974, 1978) is not regarded here as a major feature within international contracting, specifically because of the nature of the industry. This argument stems from two basic propositions concerning international construction. Firstly, as indicated in the previous chapter, the nature of the contractor's product is such that barriers to entry will not provide a significant deterrent to market entry. The mobile nature of the product, and the ease with which a competitive product may be attained (ie the 'buying in' of expertise) suggest that policies designed to maintain or erect oligopolistic barriers could easily be vaulted by any enterprise with sufficient financial resources to accumulate the necessary human capital. This implies that while financial constraints may present a general barrier to the industry, these are not a product of firm specific strategy on the part of the market leaders, who will be unable to maintain effective barriers for the reasons outlined.

Secondly, the market for international construction is greater than any national market because it encompasses several regions in which demand for construction activity is high, because of the need for extensive development. This not only means that collusion amongst market leaders would not be effective within the industry because of the costs and problems of policing the agreement, but also that the market will be too large for any enterprise to dictate policies upon the industry without some significantly powerful (and different) O advantages in relation to competitors' advantages. As we have already argued, the generation of this type of advantage is unlikely in international construction because of the nature of the industry.

On the basis of these propositions, it is suggested here that there are unlikely to be any clear market leaders that dominate policy within the industry, barriers to entry to the industry will be limited to financial constraints, and barriers to entry of foreign markets will only be effective when backed by host country government legislation. According to the works of Casson and Norman (1983) and Casson (1983) this situation will not therefore be one where oligopolistic strategy between market leaders will influence the industry. The three characteristics of international contracting as mentioned above will effectively prevent development of the industry into a true oligopoly, so that the role of oligopolistic strategy and reaction within international contracting is unlikely to be effective or an instrumental tool of competition. As such the works of Knickerbocker and Graham are potentially of little relevance in international construction.

4.3 LOCATION ADVANTAGES

The major point concerning L advantages that is related to the concept of O advantages in the eclectic theory is that L factors are immobile and must be exploited at their source. This introduces certain complications to the analysis of L specific characteristics that arise through the relationship of these advantages with the other conditions of the eclectic paradigm. While the third condition of the eclectic theory appears as a simple prerequisite for FDI to take place, it is only valid when taken in conjunction with the other conditions relating to O and I factors. Hence we may find that the advantages of a particular foreign location may themselves generate O advantages that may only be exploitable through internal organisation, or that the choice of location may be motivated by spatial imperfections (eg tariff barriers). Clearly this limits the possibility of identifying L advantages without the addition of O and I motivation for locational choice.

As with O advantages within the eclectic theory, two kinds of market imperfection may affect the location decision, structural and transactional market distortion. Structural market distortions include government intervention, productivity and cost differentials and the like, which may either encourage or discourage FDI in a certain market or set of markets. Even in the absence of

Application of MNE Theory to International Construction

these, however, if there are transactional advantages to undertaking FDI (eg the ability to exploit national tax differentials via transfer pricing, risk diversification) then international production may take place.

Given that L factors will be interrelated with structural and transactional market imperfections, Dunning suggests that choice of location will be determined according to characteristics of the home and host country, and the physical and/or 'psychic' and 'economic' distance between them. Country specific characteristics will therefore be relevant in the generation of L advantages, which is similar to the case of O advantages as given above. The major links between L factors and country specific characteristics are outlined in Table 4.2. The analysis of Lall (1980) concerning the transferability of O advantages is relevant in this context, as the immobility of L advantages determines the mobility of O factors; hence for example Table 4.2 suggests that a country whose comparative advantage rests on the abundance of natural resources is less likely to undertake FDI than a country which has a comparative advantage in, say management skills. In general this suggests that the propensity to exploit O advantages via FDI will vary according to each country's ability to generate mobile advantages, which will be a function of the home country environment.

While we may argue that the interplay of O advantages and country specific characteristics provides us with greater predictive powers concerning the exploitation of O advantages in specific areas, it should also be noted that the addition of L factors to this extended model of the eclectic theory takes the analysis further. This suggests not only that the firm will invest in foreign markets according to nationality and possible government intervention, but that the interaction between all possible O advantages (structural and transactional) and L factors may be accounted for by country specific characteristics. The consequences of this are twofold. Firstly, the nationality of the individual firm becomes even more relevant if L advantages are included so that nationality of the MNE becomes important in explaining structural and transactional O advantages and choice of overseas market, and how the two interact; and secondly the fact that country specific characteristics generates both O advantages and L advantages suggests that there is a two way relationship between the factors - hence while it has

Table 4.2: Links Between Selected Location-Specific Factors and Country-Specific Characteristics Likely to Affect Them.

1. **Production costs**
   Labour costs/productivity — Obviously greatest difference between developed and developing host countries.

   Extent to which economies of scale are possible
   Nature of production process — Size and character of markets. Factor proportions/markets in home country.

2. **Movement costs**
   Transport costs — Distance between home and host country great, therefore likely to impede countries which are of above average distance from main markets.

   Psychic distance — Where cultures, customs, language etc. different, barriers to setting up production units likely to be greater (Luostarinen, 1978) cf. Japan and US firms in Western Europe.

3. **Government intervention**
   Tariff barriers — May be country-specific in character, e.g. EEC tariffs for non-EEC investors.

   Taxation — Varies between countries. Also affected by double taxation agreements etc.

Table 4.2: continued

	General environment for foreign involvement	Political, economic, etc., ties between governments of countries.
	Incentives policies towards foreign direct investment	Again will considerably vary between host countries and sometimes home countries treated differently. Also attitudes towards outward investment may differ between home countries.
4.	Risk factors	
(a)	general propensity for foreign investment	Environment of home and all foreign countries.
(b)	geographical distribution	Cf. risk in host countries to which particular home countries most likely to invest.

Source: Dunning (1981), p.95

already been argued that certain L advantages lead to O advantages, the converse may also be true. The nature of construction in general and international construction in particular suggests that the major factor that the contractor must consider when deciding to locate overseas is level of market demand in the proposed region. Regardless of other locational factors, if demand for construction is neither experienced nor anticipated within a region or country there is no economic justification for the contractor entering that locality. This is essentially because of the nature of the final product, which will be a custom built investment good that requires a great deal of capital expenditure on the part of the client. The combination of these characteristics, plus the fact that production of the final product must take place where it is demanded for utilisation by the client, suggests that construction will be client led so that the contractor will have to follow demand rather than enter a market and stimulate demand via advertising or other tools that could be used in say, manufacturing to warrant FDI.

A consequence of this locational characteristic is that contractors may aim to search world markets for demand, but that additionally they may not enter regions where there has been a great influx of foreign contractors or where there exists a highly competent indigenous construction sector which adequately caters for the domestic construction demand. This may be added as further justification of the limited potential of a 'follow the leader' type strategy in international contracting. While the defending contractor who follows the leader in a Knickerbocker type strategy may spoil that market for the leader, the mobile nature of the enterprise and the lack of entry barriers means that the leader could feasibly enter several markets where market demand warrants it, that the follower would also have to enter for the strategy to be effective. This ultimately would prove to be both a pointless and extremely expensive policy on the part of the follower and would not necessarily restrict the leader, thus we would further argue that the nature of both O and L advantages justifies the ineffectiveness of this type of strategy for maintaining or increasing market share in international construction. Furthermore, in order to maintain demand for their services in certain key markets, contractors in the international sphere may benefit from trying to avoid each other rather than follow similar locational choices, though this clearly depends upon level of demand within that market. This may be added as a further consequence of the inability of the contractor to motivate demand.

While market demand provides the fundamental basis of locational choice, other factors are likely to influence choice of individual market given this necessary condition. According to Dunning, the essential locational factors in any industry in this respect are likely to revolve around home and host country government policy, host country political risk, and cultural and economic ties between the home and host country.

Home and host country government ties are likely to be significant in relation to the exploitation of country specific O advantages generated by government policy to the industry. This may also be reflected in cultural and economic ties (that ultimately may be a reflection of political relations) such that the greater the affinity between the home and host country in these respects, the higher the probability of success of contractors from that home country in the region. Because contractors require client led demand, it is important that the market in which they bid is not biased against them because of nationality (due to political or cultural relations), since this would inevitably lead to the contractor getting a minimal share of the market. The contractor will, therefore, in all probability seek an overseas market where there are favourable links with the home country of the contractor so that the firm is not discriminated against on nationality grounds, and may in cases enjoy a favourable environment to other nationality contractors. Clearly the influence of this depends upon how the host country values those links, such that positive links between the home and host country will only be successful where the host country clients (both public and private) respond to such factors.

In contrast to government and cultural links, the greater the level of host country political risk the less attractive may that location be to contractors, for fear of default of payment for services rendered, expropriation of assets, and even physical danger to company personnel. While this may be catered for to a certain degree, generally contractors will be reluctant to enter highly unstable markets because of these risks.

The presence of government and cultural links as a determinant of locational choice clearly illustrates the interaction of firm and home country factors with the host environment as the major criteria behind choice of individual markets of the contractor. This not only elucidates the interaction of O and L advantages, but also suggests that it is the relationship between these factors that is important to locational choice. While market demand is a necessary condition for international contracting, it is not sufficient to explain locational choice in individual markets. This choice is likely to come down to the exploitation of O advantages, in particular country specific factors that may influence choice of location according to nationality of the contractor.

4.4 INTERNALISATION ADVANTAGES

The idea that I factors are interlinked with O and L factors has clearly already been introduced by incorporating aspects of transactional costs and benefits into these components, which may be distinguished as being attributable to internal organisation. Hence the I condition of the eclectic theory suggests not only that internalisation of O advantages must take place to take advantage of L factors, but that there are significant benefits connected to these O and L factors as a product of internalisation. As already argued in Chapter 2, this negates any criticisms of the eclectic theory concerning tautology of the three conditions.

Within the context of transactional aspects of multinational involvement, I and L factors may be regarded as two perspectives of the same situation. The MNE may enter a market to take advantage of a locational factor such as relatively low national tax rates for example, but by doing so will increase the potential for transfer pricing that is only available through internal organisation. If we introduce into this argument the fact that transfer pricing will confer a transactional O advantage upon the firm, it becomes clear that any explanation of international production will be a complex interrelationship of the three conditions of the eclectic theory, and that country specific factors play an important part in determining why FDI takes place.

While some I advantages may be common across industries (eg the ability to undertake transfer pricing) there are likely to be industry specific traits that affect the propensity to internalise within an industry based upon the product and production process of that industry. Buckley and

Application of MNE Theory to International Construction

Davies (1979) give several conditions under which FDI is preferred to licensing in this context, but a more dynamic analysis that has relevance to the eclectic framework to explain industry specific internalisation is that of Buckley and Casson (1981). Given three possible modes of market servicing (exporting, licensing, and FDI) Buckley and Casson argue that the firm with a simple product (ie relatively straightforward in production in all modes) will normally enter the market via exporting, switch to licensing, and finally undertake FDI if market demand progressively increases through time. The motivation behind this switching is cost; the greater the fixed costs the more ability there is to exploit economies of scale providing the market can sustain that production level, and therefore the greater the impetus to invest in the most extensive fixed cost (and lowest variable cost) form of production. If we assume that our total cost function is given by:

$$C_i(Q) = F_i + V_i(Q) \tag{1}$$

where:

$C_i(Q)$ = Total cost of producing quantity Q by ith mode

F_i = Fixed cost of producing by ith mode

$V_i(Q)$ = Variable cost of producing quantity Q by ith mode

and the possible i modes are:

Exporting (E)
Licensing (L)
FDI (F)

then for a simple product the following relationships hold:

$$\begin{aligned} V_E > V_L > V_F \\ F_E < F_L < F_F \end{aligned} \tag{2}$$

The justification of equations (2) is straightforward. Exporting imposes relatively low fixed costs on the firm additional to its existing operating costs, relating mainly to the cost of establishing distribution outlets. Variable costs can, however be expected to be relatively high as they incorporate transport costs, tariffs, and other costs associated with export. Licensing incurs lower variable costs than exporting because the majority of variable costs associated with the physical movement of the product will be foregone, but are replaced by the cost of the licence to the licensee and the costs to the licensor of supervising the licence. Fixed costs are higher than in exporting because the licensee needs to adapt current production to incorporate the licence. In the case of FDI variable costs are clearly further reduced, although fixed costs will be higher than in licensing as a foreign production base must be set up. When taken together, as in Figure 4.1, the cost conditions show for a simple product that the firm will produce along the least cost envelope of production, ie as market size increases the firm will move from exporting to licensing and finally to FDI.

As an introduction to the notion of internalisation within international construction, it is necessary initially to illustrate the options open to the international contractor concerning the exploitation of O advantages:

1. Exporting. This involves moving personnel around markets and projects according to demand for the contractor's services. The mobile personnel may operate out of a subsidiary or the firm's headquarters, but would not usually work in those places where they are based. Construction differs from manufacturing in this respect in that in construction the mobile nature of the production base (ie skilled personnel) suggests that exporting involves the transportation of the production base to the final product until completion, which is clearly the converse of manufacturing.

2. Licensing. It is likely that licensing in international construction, if relevant, would involve the hiring out or selling of the contracting firm's name, since this embodies reputation, skills and expertise, and a guarantee of a certain level of quality, it would provide the licensee with a distinct advantage in the bidding stage and as such an obvious monetary benefit to the holder of the licence. In addition to this a firm's name is easily transferable through legal channels between enterprises, and so would be relatively straightforward to move from company to company.

3. FDI. This will involve undertaking production in a foreign country, and as such will be similar to exporting. The major difference is that in FDI the personnel are based in a permanent or near permanent subsidiary.

Application of MNE Theory to International Construction

Figure 4.1: Buckley and Casson (1981) Model for a Simple Product

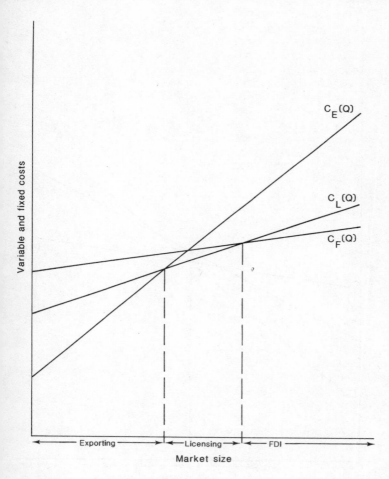

In the context of the Buckley and Casson analysis, and given the various options open to the international contractor, it is argued here that licensing will not be a feasible option in international contracting. The reasoning behind this is that the costs of licensing will be such that they do not cross the least cost envelope in Figure 4.1, so that the firm would move from exporting straight to FDI where market demand warrants it. This situation is

Application of MNE Theory to International Construction

Figure 4.2: Buckley and Casson (1981) Model for a Complex Product

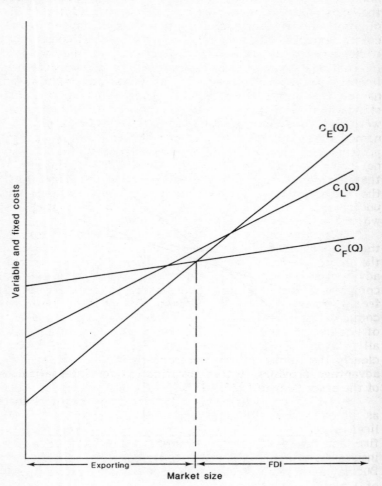

illustrated in Figure 4.2.

The higher costs of licensing that lead to the situation given in Figure 4.2 may be considered transactional costs of the external market that confer benefits upon internal organisation as suggested by Casson (1982). Ultimately, because the contractor sells a service that requires a constant product quality essential to maintain reputation (a major firm specific advantage as suggested in Section 4.2), the costs of licensing may be considered twofold in the case of international construction - potential costs and actual costs. Potential costs are those costs to the licensor that could emerge as a result of licensing. In the case of international construction these are likely to revolve around licensee underperformance. By utilising the contractor's name, the licensee effectively guarantees a minimum standard of quality as already mentioned. If the licensee were not to attain this standard, this would tarnish the name, and hence reputation, of the licensor both on present projects and future bidding situations despite the licensor not being directly responsible for the underperformance. In that this could jeopardise the licensor's whole future within the industry, it is argued here that the possibility of underperformance would provide the licensor with what would be considered an unacceptable risk.

This argument is more emphatically stated if we argue that although the name of the firm is easily transferable, the factors that make the name a major firm specific advantage are not. The contractor's name represents a complex mixture of intangible assets that would not be freely transferable under license. Given that the more complex and intangible the asset the higher the probability of licensee underperformance because of failure to identify all relevant aspects of the advantage (Teece, 1983) then clearly the nature of the contractor's major firm specific advantage provides further justification for internalisation of the asset (Magee 1976, 1977).

Actual costs of licensing will arise from potential costs as given above. To prevent underperformance by the licensee, a contractor would have to search for a suitable firm and police the licence via extensive legal agreements, insurance, and quality control supervision. In terms of the Buckley and Casson model when coupled with potential costs of licensing, this would raise both the fixed and variable costs of licensing above the least cost envelope, as argued above and illustrated in Figure 4.2. Based upon this argument and the model, it is therefore to be expected that in international contracting there will be no or little external exploitation of O advantages.

Although in many cases exporting and FDI will be substitutes, the similar nature of the two modes of market servicing within international construction may lead to simultaneous use of exporting and FDI in the same market under special circumstances. Because of the mobile nature of the contractor's services, exporting and FDI will only be separable by the amount of time the personnel are within a market, and so will be easily interchangeable. This may be regarded as a consequence of the fact that in construction the production process will be located where the final product is produced, and not vice versa as in manufacturing. There are likely to be two major cases where simultaneous use of exporting and FDI will be used:

1. Where demand is spasmodic. Using both exporting and FDI may be the norm in markets where contractors experience widely fluctuating demand. By such a policy the firm may utilise its personnel resources worldwide and be able to compete for major contracts but still maintain local bases in the major markets for administrative work and small scale construction projects. Demand for construction services of the firm is likely to be most spasmodic in markets where the size of projects is large and the number of competitors at the bidding stage high. In this situation chances of winning a project are significantly reduced, making it impossible to accurately forecast future demand for the firm's services. In markets where construction is smaller scale, there are fewer competitors and stable demand, the contractor will use either exporting or FDI depending upon the level of market demand and locational criteria.

2. Where risk is high. The contractor is likely to aim to minimise the firm's fixed assets within a politically unstable country, but at the same time maintain a local presence for bidding and/or administrative purposes. Clearly exporting and FDI together provide a solution to this problem; by sustaining a local office and 'importing' personnel into the country at times of high demand for the contractor's services, the contractor will ensure a minimum risk exposure, that will generally not put the firm's personnel at risk since they will be the most mobile of the firm's assets.

While we have seen above that the international contractor is likely to prefer internalisation to the external market as a means of maintaining product quality, the nature of the contractor's product (that is intermediate) suggests that the firm is faced with more alternatives than simply exporting, licensing, or FDI in the foreign market. Following Casson (1985b) we distinguish several possibilities

Application of MNE Theory to International Construction

of equity and non-equity involvement of the firm:

1. FDI, where outright control of the firm's assets remain with the contractor so that direct investment abroad is the only way of foreign involvement.

2. Joint venture, where control of production is shared between two partners, usually on a 50/50 basis of equity, so that risk is likewise shared (Dunning and Cantwell, 1982).

3. Industrial cooperation agreement. This is essentially a joint venture where the parent firm's equity involvement is only short term, as in turnkey projects.

4. Sub-contracting. Under sub-contracting, production is delegated to another firm and the parent (or main contractor) markets the final product under its own name.

5. Licensing has already been discussed, but serves here as a market alternative to FDI in that the licensor has a non-equity stake in production.

6. Sales franchising. Under sales franchising the marketing rather than production is delegated to another firm who sells the product according to the franchise agreement.

7. Management contracting. Management contracting may be considered a licensing or sub-contracting arrangement whereby the contractor provides management for the client to produce the product through training and supervision (Gabriel, 1967). Typically the management contract may be a form of industrial cooperation as outlined above.

Casson (1985b) suggests that given the possible alternatives, choice of equity or non-equity participation will depend upon two factors; the level of risk between the parties involved (as reflected in ownership structure), and the allocation of management responsibility as reflected in the structure of control. Since each possible alternative will have different advantages and disadvantages relating to the internal hierarchy and external market, the asset owner must decide the best alternative given the location of use and nature of the advantage. Following on from the analysis of Casson, Table 4.3 illustrates what are considered here to be the major advantages and disadvantages necessary for the decision between market and hierarchy (within the forms given above) in international construction. An arrangement that contributes to the minimisation of a strategic problem is marked with a plus (+) sign, while a negative (-) sign denotes that the arrangement exacerbates

Table 4.3: Strategic Factors Affecting the Choice of Contractual Arrangements within International Construction

Strategic Issue/ Contractual Arrangement	FDI	Joint Venture	Turnkey	Management Contract	Sub Contract	Licensing	Sales Franchise
Maintain quality control during production	+	+/0	+	+	0/-	-	+
Preserve advantage	+	+/-	0	0	0	-	-
Acquire capital	-	+	+	+	+	+	+
Obtain knowledge of local production environment	-	+	+	+	+/0	0	-
Obtain knowledge of local marketing environment	-	+	+	+	+/0	0	+
Operational integration of production & marketing	+	+	+	-	-	+	-
Repatriate profits by transfer pricing	+	+/0	-	-	-	-	-

Table 4.3: continued

Strategic Issue/ Contractual Arrangement	FDI	Joint Venture	Turnkey	Management Contract	Sub Contract	Licensing	Sales Franchise
Avoid management conflict	+	-	-	+	-	0	0
Avoid political risk	-	+	+	+	+	+	0
Maintain goodwill of host government	-	+*	+	+	+*	0	0
Avoid communications problems	+	-	-	-	-	+	0
Need for specialist services outside the firm	-	+	0	0	+	+	+

* Where partner is indigenous firm

Source, modified from Casson, 1985b

the problem. A zero (0) sign suggests that the arrangement makes a modest contribution to resolution of the problem. The strategic issues given are considered obvious in terms of this discussion and therefore need no exposition.

Although Table 4.3 illustrates the positive and negative factors connected with each contractual arrangement given, it should be noted that certain strategic issues will be more relevant than others, depending upon the advantage in question, and to a lesser extent the market situation. In terms of the analysis here, and the Buckley and Casson (1981) analysis given above, possibly the most strategic issues in Table 4.3 for the international contractor are the maintenance of quality control and the preservation of competitive advantage. Because of this, some forms of alternative in the table will not be available in the industry because of the nature of quality control:

> Quality control is most effective when administered through close supervision of the production process, since this avoids inferior quality items being produced in the first place. The key tactic here is to ensure that the parent firm's proprietary advantage in production is effectively transferred to the local production unit, and that staff trained by the parent firm supervise the production ... Outright control, joint ventures, industrial cooperation agreements and management contracts normally provide this assurance, whereas the other contractual arrangements do not. (Casson p.17, 1985b)

This provides further justification for the lack of licensing in international construction, and additionally suggests that sales franchising is not likely in the industry while subcontracting will only be possible where quality control and asset advantages of the contractor may be protected. This will be taken up below in relation to vertical and horizontal integration.

Of the options that are open to the international contractor, the choice of alternative may depend upon the situation in which the advantage is likely to be used; for example FDI may be the preferred alternative in that it is the most effective way to maintain quality control and preserve the competitive advantage of the contractor, but where there is political risk in a market joint venture might be chosen as a more suitable option to lessen the firm's exposure (Vernon, 1983). Given Table 4.3, we can predict where each of the potential alternatives may be the desired form of involvement given the local environment and demand for the contractor. This is summarised in Table 4.4.

Table 4.4 shows quite clearly that each possible alternative is likely to have both benefits and costs, but it is also significant because the table illustrates an interesting, and important point about the nature of involvement by international contractors. Although it is implicitly assumed in MNE literature that ownership stake and internalisation are closely correlated so that 100% equity stake is necessary for complete control, the similarity between the various options as given in Table 4.4 suggests that in international contracting this is not necessarily the case. In the table it is apparent that a non-equity alternative such as management contracting for example will contain characteristics of control of production of the final product that would only normally be associated with FDI in other industries. Through the management contract the contractor can effectively control and supervise all on site work despite having no equity share in the project. In this sense, internalisation in international construction may be extended beyond basic 100% ownership of production to apply whenever a firm of one nationality has control of production in another country other than 'spot' or arm's length off the shelf transaction. Hence joint venture, turnkey, management, and sub-contracting may be considered within the internal organisation of the contractor, whereas in manufacturing these could be regarded as a form of externalisation.

It should be noted that the prevalence of this notion of internalisation within international contracting may be explained by the nature of firm specific O advantages within the industry. The success of an international contractor to coordinate inputs and supervise production to construct a superior product to indigenous contractors in LDC regions is ultimately due to the fact that the contractor relies upon human rather than physical capital to provide a particular service. Because of the nature of this capital, the contractor may use the advantage under contract without fear of disclosure or imitation (that could effectively destroy the O advantage if the advantage was superior technology embodied in physical capital), and so still earn an economic rent because of legal restrictions and the ability of the firm to protect its name (that will reflect the

Table 4.4: Options Open to the International Contractor in Foreign Markets

Advantage of Option	Disadvantage of Option	Situation Where Likely
F.D.I. Allows quality maintenance and preservation of advantage without fear of loss; gives rise to internal benefits occurring to interal integration and transfer pricing that are likely to be greater than in any alternative.	Restricts capital to firm, funds and skills to those available within the company; does not allow firm to know all about market straight away, and is likely to arouse government suspicions and be susceptible to political risk or xenophobia on nationalistic grounds.	Where quality control and advantage protection are at a premium, detailed knowledge of the local market is not required, political risk is low (and host government protectionism) and the firm can fully cater for capital and skill demands.
Joint Venture Allows quality control and maintenance of advantage to a point; does not restrict operations by capital or skill requirements and may have local knowledge through joint	Firm must maintain strict quality control to prevent default by partner that would affect the contractor's product; joint venture may put together incompatible able partners	Where capital, specialist skills or local information is required for work; where political risk is high, to prevent exposure (Vernon, 1983); where the host government

118

Table 4.4: continued

Advantage of Option	Disadvantage of Option	Situation Where Likely
venture. Less exposed to risk than FDI, particularly where in joint venture with indigenous contractor.	that makes both inefficient. Joint venture may prevent internal benefits of FDI where the partner is an indigenous contractor.	follows protectionist policies and regulations that require joint venture with local contractors.
Turnkey Contract Allows quality control and enables the firm to acquire capital, and learn about host market. Lessens political risk exposure of the firm to host government protectionist regulations and bias.	Is likely to lead to partial loss of advantage through training of nationals; prevents internal benefits of FDI and may actually increase risk exposure of the firm in cases of revolution or war. Possibly increases conflict between contractor and indigenous ideals and methods.	Where host government stresses the need for equity participation in building and requires training of nationals; where capital is required and local knowledge of market is sought (although these may be secondary to government policy).
Management Contract Allows most of incentives of turnkey contract as stressed above, but additionally faces less risk exposure in any situation.* Is likely to	May lead to loss of advantage as host nationals learn the operations of the contractor, and will not allow internal exploit-	Where host government stresses the need for management contracting and training of nationals; where capital is required and local knowledge of

Table 4.4: continued

Advantage of Option	Disadvantage of Option	Situation Where Likely
be favoured by host government.	ation as in FDI. Additionally may lead to communications problems between management contractor and indigenous firms.	the market is sought. Where the contractor wishes to minimise assets in a market because of risky environment.
Sub-Contract Minimises risk exposure of the main contractor; allows additional services and capital for the contractor to work with. May help avoid host government risk and improve goodwill where sub-contractor is indigenous and increase local knowledge.	Leads to the possibility of under-performance by the sub-contractor that will be reflected in the main contractor's product; prevents internal benefits of FDI and leads to communication problems between main contractor and sub-contractor.	Where the main contractor can maintain quality control of sub-contractor; where specialist skills or capital are required or where the host government favours or demands local involvement; where risk exposure is likely to be high without sub-contracting.

* In terms of fixed assets and commitments.

Source: Table 4.3 and Casson, 1985b

Application of MNE Theory to International Construction

competence of the contractor's personnel).

There may however also be transactions costs associated with contract; by undertaking a contract type arrangement the contractor will lose benefits of internal organisation that are exploitable through FDI, such as the ability to repatriate profits or maintain market research in that area (since under contract, the firm's involvement finishes on completion of the contract). However this may be catered for, as suggested in the Buckley and Casson model mentioned above, because the international contractor does not have to restrict choice to either FDI or contract (that may be considered forms of exporting in that personnel are moved into an area when demanded), but may simultaneously undertake both. The ability and advantages arising from the 'mix and marry' of these alternatives is generated by the separation of production and marketing responsibilities within the company (Casson, 1985b). Within international contracting it is argued here that the contractor is likely to utilise non-equity or contractual form of involvement in production in overseas regions to avoid and minimise risks to the firm in any specific project, but additionally will operate a subsidiary through FDI where the firm wishes to establish a marketing operation to obtain local work (through reputation and name of the firm). While the benefits of contractual arrangement have already been discussed, it should be noted that the contractor will have to internalise the name and reputation of the firm (that are the basic components of firm specific marketing in this context) by direct control to reduce uncertainty and possible misrepresentation that could be experienced in the external market. This point is clearly highlighted by the disadvantages of sales franchising in international construction as given in Table 4.3. In general terms therefore it appears likely that the benefits of the contractual relationship in production cannot adequately be provided in marketing of the firm overseas so that internalisation of marketing will demand an ownership stake rather than a contractual relationship. While this may be regarded as a function of the intangibility of the assets required for marketing (and therefore justification for Magee's appropriability approach (1977, 1978)), it also provides further justification for the simultaneous use of exporting and FDI in certain areas as suggested by the Buckley and Casson (1981) model when applied to international construction.

In terms of theoretical predictions therefore we would expect that subsidiary offices will be set up where overall market demand is considered attractive in that market (to continually research and market the company in that country), but that construction activity will generally involve exported personnel where risk is high and/or demand for the firm's services is sporadic, as a means of minimising the contractor's exposure to risk and maximising resource use across markets. In cases where the host country requires joint venture with a local partner in order to bid for contracts, the contractor may still maintain a contractual approach to production, but is likely to market in that country as part of the joint venture to minimise risk to the firm's name and gain favour with local clients and the host government. This may essentially be considered a special form of FDI in marketing that in this respect is undertaken for different reasons to joint venture on a specific project (which is more in line with a contractual type involvement), and is likely to predominate where protectionist or nationalist governments are in office in host countries.

The notion that product quality provides the motivation behind internalisation of the firm's assets in international construction leads to further analysis of the nature of the industry that may be derived from the benefits of internal organisation. Casson (1986) for example has suggested that the extent of vertical integration (VI) in any industry may be determined by the potential level of exploitation of the internal organisation according to technical, market power dynamic or fiscal factors related to the firm and industry that may have positive or negative influence. In terms of international contracting, the level of VI is likely to depend both upon the ability of the contractor to ensure product quality and the full utilisation of the resources available to the contractor.

Within international construction, backward integration would take the form of pre-construction activities that generally relate to manufacture of materials for construction, and services necessary for construction to take place (eg consultancy, feasibility studies, design, and financing arrangements). Forward integration may involve movement into activities carried out after construction (eg some types of sub-contracting, maintenance of final product). Casson (1985c) argues that given the nature of international construction, VI is only likely to take place where the contractor relies heavily upon a specific product

Application of MNE Theory to International Construction

used within construction and there is a benefit to internalisation; in the case of certain building materials such as cement and aggregates for example, by internalising production of these materials the contractor is able both to ensure quality of that material and guarantee regular supplies of the materials at an internalised price. Similarly, where the contractor requires a specialist service at a specific period of time in the construction period, internalisation may be the only risk free way of maintaining quality and construction activity if a sub-contractor cannot be found to carry out the work to the specified requirements. In both cases the motivation for internalisation remains product quality guarantee where the external market is either inefficient or missing, although other benefits of internalisation may also occur. In the case of internalisation of materials for example the contractor benefits from the ability to obtain materials at cost price, and therefore may be cheaper than competitors (or may undertake price regulation based upon internal pricing).

VI is only likely, however, where the product forms a significant part of the contractor's business, on the grounds of maintaining efficiency of the contractor. For example while VI into maintenance may be considered a feasible option because it is likely to offset demand fluctuations for the firm's services (Casson 1985c), in general resources that have alternate uses in other industries will not be internalised because this would create inefficiency both to the resource and the internaliser through underuse of that resource when demand is low for the contractor's services. For this reason architects or providers of financial services for example are not likely to form part of VI strategy for the majority of contractors, though where they do the internalised resource will maintain links with other industries through the hiring out of skills when not required for construction services by the parent. This may be regarded as partial conglomeration and partial VI though it should be noted that it will probably increase the demand upon management who will need to coordinate both the main construction activity and the subsidiary involvement in other industries and therefore may not be acceptable to the contractor.

While the benefits of VI to the international enterprise in general are potentially numerous, including the possibility of transfer pricing and price regulation setting (Casson 1986), in international contracting the benefits of integration into materials necessary for construction may be reduced or possibly eliminated altogether. The motivation for VI suggests that the contractor integrates to maintain stability and therefore the firm will attempt, where possible, to integrate in the home market to minimise risks and costs associated with international involvement. The contracting firm therefore faces the risk of nationalistic pressure in overseas markets to utilise host country produced materials. Since this is likely to be the case in many LDC markets where great emphasis is placed upon the development of an indigenous materials sector for infrastructure improvement, the scope for benefits may be minimal. However, the use of internalised plant operations and various service departments within the firm will generally not be affected by such protectionist policies, and may provide economies of scale that are considered internal benefits to VI.

If we consider that the incidence of horizontal integration (HI) will be motivated by factors similar to VI, then internalisation provides a means for diversification of the firm that in this sense is both a structural and transactional O advantage as outlined in section 4.2. Since HI involves the integration of specialist construction services into the firm that is based upon expertise and experience of a person or team, essentially HI involves the employment of personnel in specific areas of construction that may additionally be utilised via sub-contract or temporary employment. In terms of internalisation theory Casson (1985c) argues that 'The benefits to the main contractor of offering regular employment are greatest when the fixed components of transactions costs are highest.' (p.8, 1985c).

This may be reflected in analysis similar to that given above for VI. The contractor is likely to internalise only those personnel that do not have alternate uses in other industries (for a given skill), and contribute significantly to the construction work of the contractor in total, for reasons of efficiency. It is therefore likely that the contractor will internalise a management team so that confidence communication, and individual skills may be built up between the contractor and various members of the team that could not be possible to the same degree in the external market (Penrose, 1956). On this basis Casson suggests '.. managers engaged in project planning are likely to be offered permanent employment' (p.9, 1985c) while highly specialised or unskilled personnel will be hired temporarily to minimise wastage of resources within the firm.

To maintain efficiency within the firm, the contractor will have to diversify management because of the cyclical nature of demand for the firm's product in the industry. Diversification may take place via different locations, type of client, technology demanded, or size and complexity of project, depending upon the firm's strengths and weaknesses. It should be noted that the diverse use of personnel according to location and type of technology demanded are likely to be the major features of HI in international construction since these are potentially the determinants of both type of client and size and complexity of project, as outlined in Chapter 3. In this context, the Buckley and Casson (1981) model may be a valid explanation of the interaction of these factors in a dynamic setting in the industry. Specialist skills within the firm may not be needed throughout the life of the construction project (eg design engineer), and therefore it is in the contractor's interests to move personnel from project to project in various locations, and to maintain fixed subsidiaries in specific locations for minor construction works and marketing only, as already discussed. The diversity of personnel use through contractual and FDI methods in HI may be considered a benefit of the internal organisation in that it leads to benefits to the contractor that will not be exploitable in the external market.

HI is likely to take place in international construction for two reasons:

1. Product quality. If we argue that the product of the sub-contractor is likely to affect the product of the main contractor in construction (since it will generally be the case that the main contractor is solely accountable for the final product), then clearly there may be impetus for the contractor to consider the internalisation of certain skills where they may pose such a threat in the external market. Where the skill forms a major part of the contractor's product, the less satisfactory the external market the more likely is internalisation of that skill to remove uncertainty concerning the supply and quality of the skill that would be reflected in the quality, and hence reputation, of the main contractor. While this argument is similar to that given above for VI, it should be noted that HI in this sense may be considered as a means of minimising transactional costs that would be present in the external market.

2. Benefits of diversity. By diversifying into various sub-industries by HI, the contractor will experience several benefits that may be regarded as transactional advantages in the eclectic framework.
   a. Demand will be more stable and less prone to cyclical fluctuation as a result of diversification as the contractor may locate resources according to demand in various locations and specialist markets. Additionally, as suggested above, management resources of the firm will be fully utilised and therefore efficiently exploited.
   b. The contractor may in any bid situation offer a wider range of services and thereby differentiate the firm from others who would have to sub-contract specialist skills. This may be seen as an advantage in that the contractor differentiates his bid, but also because within the internal organisation price fixing may be more efficient (and less exposed to external investigation) than in the external market.
   c. By operating in several markets the contractor's reputation and name will be enhanced in all markets so that chances of bidding success are greater overall.

Chapter Five

EMPIRICAL ANALYSIS OF OWNERSHIP ADVANTAGES

5.1 METHODOLOGY OF THE RESEARCH

The majority of the results presented in this and the next chapter are from detailed discussions with executives of major contracting companies that are involved in overseas markets. The discussions were based upon a specific set of questions which were presented in questionnaire form, which was considered the most effective way of obtaining information because it allowed interviewees to express their opinions and relate details of the industry in general and their company's performance in particular, while at the same time enabling statistical analysis of the answers to identify common features between participant contractors. This was backed by additional interviews with people connected with the industry which formed much of the background upon which the empirical study was based.

Because it was felt that the study should reflect the general situation in the industry, choice of contractors approached was not limited to a specific size of contracting firm or construction speciality. Criteria for choice of contractors was basically that the contractor had to be involved in overseas construction at the time of interview, and that the firm had an office in the UK. (1) Contracting companies' names and addresses were taken from several publications and organisations, including the Export Group for the Constructional Industries directory of UK contractors, Engineering News-Record's annual list of the top 250 international contractors (and subsequent identification of the enterprise within the Kompass directory of registered firms in the UK), the Times 1000 and various embassies and trade councils. Where there was evidence that the company does undertake international construction (usually from the ENR top 250 international contractors list, and from discussions with people connected with the industry), a letter was sent to the managing director of the company outlining the research and requesting an interview with an executive of the firm involved with overseas construction. Table 5.1 illustrates the response of the contractors approached; of the 64 contractors that were sent details of the research, 20 companies agreed to participate while 18 declined. 26 did not respond at all, despite several attempts to contact the director in charge.

Table 5.1: Response Rates of Participants in Survey

Contractors willing to be interviewed	20
Contractors declining to be interviewed	18
No response	26
Total companies approached:	64

Table 5.2 summarises the sample breakdown. In terms of nationality, UK contractors were the largest national group in the survey representing 55% of the total. American contractors were the second largest group comprising 30%, while the remainder of the survey were of European and Far Eastern origin.

Table 5.2: Breakdown of Samples

UK contractor head office	11
Foreign contractor operating in UK	9
Total companies in sample:	20

Although quantitative information was not available on 6 of the firms in the survey because they were either private companies or part of a large conglomerate group (and therefore not separately identifiable from company reports) firm accounts and reports, and various published articles on the remaining 14 give a fair representation of the significance of the participants within the international

Empirical Analysis of Ownership Advantages

construction industry. The major factors are summarised below for the year 1982, which was chosen because of the relatively high level of published material about the contractors in this year.

1. Level of foreign awards for the 14 participants stood at £6.61 billion in 1982. If we consider that this represents approximately 3.5 times the total value of overseas work by the British construction industry in the same year, then clearly the survey respondents form a significant representation of the international construction industry.

2. Total awards of the 14, which includes domestic work, was £11.98 billion in 1982.

3. If level of foreign awards is taken as a proxy for size of the enterprise in the international market, then firms in the survey ranged from the very small to the very large. In 1982 the smallest contractor of the 14 analysed contributed 0.1% of total foreign awards, while the largest contractor involved was responsible for approximately 35% of foreign awards. Excluding these two extremes, average size of the contractor was £355 million or approximately 8.3% of foreign awards.

We can argue that these statistics together with the results of the interviews give a fair representation of the contemporary environment within the industry.

5.2 OWNERSHIP ADVANTAGES

Given the framework of the eclectic theory, and the theoretical analysis presented in Chapter 4, empirical analysis of ownership advantages is separated here into firm specific and country specific factors to separate micro and macro influences, although clearly the two may be related.

5.3 FIRM SPECIFIC FACTORS

5.3.1 **Product differentiation within international construction**

Within the survey, contractors were generally asked to relate questions on the questionnaire to their experiences in the Middle Eastern market, on the premise that this is the market where demand for construction works has been

Table 5.3: Perceived Competitive Advantages/Disadvantages of Contractors in Survey with Respect to Indigenous Arab Contractors in the Middle East

Advantages	%	Disadvantages	%
Technical knowledge	57	Protectionist Arab government policies and attitudes	62
Experience of international operations	57	Lower cost Arab contractors	48
Management expertise	38	Knowledge of region and buyers	29
Reputation	28	Client bids in favour of Arab firms	19
Financial resource of the firm	19	Legal stipulations over joint ventures	9
Size of firm	14	Taxation bids in favour of Arab firms	5
Availability of technical resources	9		
Home government assistance	5		
Demand for home country consultants	5		
Lower costs than Arab firms	5		

Source: Field Work

greatest, and therefore competition may have developed in the context of O advantages in this market more than in any other region. Those firms in the survey that had little experience of the Middle East region related the questions to their major overseas market, though this approach was only necessary for two of the respondents.

As a means of introducing how contractors compete in overseas markets, participants were asked to list what they considered were their major advantages and disadvantages with respect to indigenous contractors in the Middle East. The purpose of the question was twofold:

1. To see how relevant the possession of O advantages is with respect to indigenous enterprises (which is of importance with respect to Hymer's work);
2. To determine how contractors perceive their competitive strengths and weaknesses.

The results are summarised in Table 5.3. The table shows that the most frequently cited perceived advantages were technical knowledge not acquired by Arab contractors (given by 57% of the respondents), experience of overseas operations (also 57%), management expertise (38%), and reputation (28%). Lesser cited advantages include financial resources, size of the firm, and access to technical resources (that may all be considered as advantages of size of the company). Home government assistance, demand for home country consultants and lower costs which are included in Table 5.3 cannot be classed as significant to the survey as a whole. Disadvantages of the contractors against local firms tended to revolve around protectionist policies and local bias in the market region in favour of Arab contractors, although the lower cost of indigenous firms and better knowledge of the region and buyers were given as highly relevant advantages of local firms.

To compare how contractors perceive competition with other international contractors to be different from competition with indigenous construction firms, participants were asked to list advantages and disadvantages with respect to other international contractors in the Middle East. The situation is illustrated in Table 5.4. The table shows that while reputation was the most commonly cited advantage (given by 28%), technical superiority or equivalence was also considered an important advantage. It is interesting to note in Table 5.4 that while the incidence

Table 5.4: Perceived Competitive Advantages/Disadvantages of Contractors in Survey with Respect to Other International Contractors

Advantages	%	Disadvantages	%
Reputation	28	Relative level of home government support	52
Technical superiority	24	Foreign companies' cheaper costs	48
Technical equivalence	24	Foreign access to cheap project finance	33
Management expertise	19	Low home market demand	9
Demand for home country consultants	19	Lower level of coordination than other nationality contractors	9
Experience of international operations	14	Home country materials supplies quality	5
Size of firm	14	Language problems	5
Lower price than competitors	14	Size of firm	5
Language similarities to markets	9	Technological ability	9
Historical links to market	9		
Political links to markets	5		
Good home financial market	5		
Industrial links with market	5		
Firms specific product	5		

Source: Ibid.

Empirical Analysis of Ownership Advantages

of management expertise has fallen as a perceived O advantage in relation to Table 5.3 (although it remains a significant advantage), demand for home country consultants and other country specific advantages tend to be more prominent, eg language and historic links. Size of firm remains at the same level as in Table 5.3. Non-significant advantages in the table include political links, a good home country financial market, industrial links between countries and the ability of the firm to produce a service which is unique in the market. It should be noted that with the exception of the last factor product these advantages relate to country rather than firm specific characteristics and therefore are treated in more detail in the 'country specific' section of this chapter.

Disadvantages with respect to foreign competitors were in the main isolated to country specific competition rather than individual O disadvantages. Relative level of home government support, foreign companies' ability to exploit cheaper costs, and foreign access to cheap project finance as the most commonly cited disadvantage in Table 5.4 can all be traced to country specific determinants in this context, and therefore will also be treated in more depth in the section on country specific advantages. Possibly the only firm specific factors cited as major disadvantages were size of the firm and limited technological ability of the contractor, though from Table 5.4 we can see that the significance of these to the survey as a whole are clearly limited.

The relevance of Tables 5.3 and 5.4 to O advantages in international contracting may be summarised by two points:

1. Table 5.3 suggests that the contractor entering the Middle Eastern market is liable to face both bias against the firm and a disadvantage in relation to contractors already in the market with knowledge of the local market environment. It therefore seems likely that the firm must have some initial advantage in order to enter and compete in the region. This is clearly in line with the analysis of Hymer (1976), and suggests justification both for the first condition of Dunning's eclectic paradigm and also the need to differentiate the firm in overseas markets in order to compete.

2. In international construction there is likely to be a product cycle type development which influences the relative nature and perception of O advantages. Because Arab contractors have evolved only in the past ten years

Table 5.5: Services Offered by Participants in Survey

%

	Always 5	Very Frequent 4	Often 3	Rarely 2	Never 1	Mean $\bar{x}$
Reputation	55	25	5	15	0	4.2
Cheaper price than others	45	15	15	25	0	3.8
Training of host nationals	10	65	10	10	5	3.6
Turnkey agreements	0	60	30	10	0	3.5
Faster completion than others	14	19	48	19	0	3.4
Locally bought supplies	16	31	42	10	0	3.3
Profit repatriation agreements	42	10	21	10	16	3.3
Foreign labour agreements	10	37	37	0	16	3.1
Subsidiary joint ownership	5	37	37	21	0	3.1
Contractor arranged finance	0	42	31	26	0	3.0
Payment in foreign currency accepted (not £s or $s)	0	37	42	21	0	3.0
Higher quality than others	20	25	10	30	15	3.0
Maximum local firm involvement	10	37	16	26	10	2.9
Local labour agreement	10	26	21	26	16	2.7
Project maintenance	17	17	17	28	22	2.5
Countertrade and host currency payment	0	17	39	33	11	2.3
Cheap access to home country products	0	10	21	37	31	2.0

Source: Ibid.

Empirical Analysis of Ownership Advantages

(mainly through government inception and support) the product they offer is relatively new and therefore not only susceptible to the failings and idiosyncracies of a new product, but is also likely to be only partially developed. This is in marked contrast to the product offered by an international contractor which will have developed through a learning curve along the lines of Vernon's analysis (1966, 1971), which in the case of international construction is likely to have involved the accumulation of a high level of management expertise and construction skills. It is therefore likely that where the international contractor competes with an Arab enterprise, there will be a significant difference in the level of technical skills offered (as suggested in Table 5.3) so that this forms the international contractor's major O advantage. However where the international contractor competes with another foreign contractor with similar expertise, the production process is not likely to be significantly different so that the contractor has to rely upon product differentiation as a means of distinguishing the firm's product (Dunning and Norman, 1983). This is the case in Table 5.4 where the major O advantage stressed by respondents is reputation. This may be considered product differentiation in that the contractor differentiates the services offered not only via technical skills, but additionally by resources of the firm and past experience (which will be embodied in reputation) which highlight the quality of the contractor's product even when in competition with firms offering similar services. The implication of this (which may be attributed to Vernon's product cycle theory) is that competition between international contractors is likely to be characterised by product differentiation which relies upon the exploitation of firm specific factors (which embodies technical expertise and other firm specific differentiable factors), that provide the basis of the reputation of the enterprise.

To analyse the empirical relevance of product differentiation within the industry, participants in the survey were given a list of services that the contractor could potentially offer as an explicit marketing tool in the bid situation, and asked to rank these from 1 to 5, with 1 corresponding to the firm never offering the service and 5 corresponding with the firm always offering the service. Clearly the more the contractor regards a specific factor as being instrumental in the winning of a bid, the more likely the firm is to incorporate that service in all bids. Table 5.5

Table 5.6: Services Demanded by Middle Eastern Clients

%

	Always 5	Very Frequent 4	Often 3	Rarely 2	Never 1	Mean $\bar{x}$
Cheaper price than others	50	33	17	0	0	3.9
Reputation	39	39	17	5	0	3.7
Joint ownership of subsidiary	22	33	44	0	0	3.4
Maximum local firm involvement	10	58	21	0	10	3.4
Training of host nationals	11	67	11	5	5	3.3
Locally bought supplies	5	67	22	5	0	3.3
Foreign labour agreement	19	37	37	0	6	3.3
Turnkey agreements	6	41	47	6	0	2.9
Faster completion than others	6	35	41	18	0	2.9
Project maintenance	22	11	33	22	11	2.8
Contractor arranged finance	6	25	50	19	0	2.8
Local labour usage agreements	12	12	44	25	6	2.5
Payment in foreign currency	0	25	56	19	0	2.4
Higher quality than others	12	12	29	29	18	2.4
Countertrade and host currency payment	0	18	47	23	12	2.3
Repatriation of profit agreements	13	13	20	13	40	2.3
Access to cheap home country products of contractor	0	12	12	50	25	1.8
						1.7

Source: Ibid.

Empirical Analysis of Ownership Advantages

illustrates the response given by the participants in the survey, with the mean figure showing the overall importance of that factor. The higher the relative mean figure, the more important the service is considered.

Contractors were additionally asked to use the same list of services and method of ranking to outline what they considered Middle Eastern clients demanded most in the bid situation, to see how coordinated the demand of the client and the supply of the international contractor are. This is illustrated in Table 5.6.

The most obvious point in Table 5.5 is that while there is quite a wide diversity of services included in the bidding stage by contractors, there are common factors which are clearly considered more important than others. The most obvious of these (as would be expected from the analysis of Chapters 3 and 4) are reputation and the ability of the contractor to offer a cheaper price than competitors. The table is interesting in this context not only because reputation was more commonly stressed than cheaper price as an important factor of bidding (and would therefore suggest that the contractor's major firm specific advantage in the bidding stage is its reputation, and hence name), but also because product differentiation via other means appear to be fairly common within the industry. Training of host nationals, turnkey projects, and the ability to construct projects faster than competitors were frequently offered in the bid stage, while protectionist factors such as locally bought supplies, subsidiary joint ownership (with a local partner) and the like were often included in the bid, however the influence of country specific locational characteristics are likely to be relevant in this respect.

Financial incentives such as making provision for the client to pay in foreign currency (to minimise exchange risk for the client), and contractor arranged project finance were considered by many in the survey to be too generalised to discuss without reference to a particular market. It was argued that the demand for these types of services would in the main come from the lesser affluent countries in the region. Hence while most contractors suggested that they would provide financial services in certain cases where required, within the Middle East the demand was likely to be variable according to the national market.

In the table the factors rarely offered as incentives in the Middle Eastern market were provision to utilise local labour, project maintenance agreements, countertrade

137

Empirical Analysis of Ownership Advantages

agreements, (ie the payment for contractor's services in the form of raw or semi processed materials), and the access to cheap home country materials of the contractor. Due to the fact that there is a small local labour sector within most Middle Eastern countries, a growing indigenous construction industry with the ability to carry out maintenance, and extensive foreign exchange reserves within the dominant OPEC countries in the region, countertrade in the past has been largely disregarded. These results were therefore anticipated as a result of the interaction of O and L factors, though in regions or markets where these characteristics are not found the provision of the services mentioned could play a more important role in bidding.

Table 5.6 shows a similar situation as Table 5.5 in that reputation and cheaper price appear as the most demanded factors in the bid situation by clients, although the importance of the two has been reversed so that price becomes the dominant issue. The reasoning behind the difference is somewhat obvious from Chapter 4. Whereas the contractor is likely to consider reputation as the most important factor in bidding because it distinguishes the firm from all others with similar skills (as we have argued above with reference to the product cycle), the client's motivation is to find the least cost contractor in the bid and therefore the client places greater emphasis upon price. However, the significance of reputation in Table 5.6 suggests that additionally the client is likely to consider name of the firm as an important aspect of competition in the industry so that product differentiation may be justified as a major element of competition.

Additional factors of importance to clients stressed by the survey respondents included protectionist attitudes (which are a function of country specific locational factors) so that joint ownership of subsidiary with local partner, maximum local firm involvement, training of host country nationals, and the provision of locally bought supplies were all regarded attractive incentives to prospective clients. This may be regarded as an aspect of local bias by the host government as suggested in both Table 5.5 and Table 5.3, which provides further justification of the need for competitive O advantages to compete successfully in the region.

While the need for foreign labour agreements remained fairly similar between Tables 5.5 and 5.6 (which are necessary because of the lack of local labour in the region)

Empirical Analysis of Ownership Advantages

there were certain divergences in the tables between the frequency of demand and supply of turnkey projects, faster completion than competitors, payment in foreign currencies, and especially repatriation of profits by contractors. This may generally be considered as the conflict between contractor and client objectives and motives, and at times a lack of communication between the two. It is sufficient here to note that these deficiencies exist.

Lesser demanded services by clients in Table 5.6 included payment in foreign currency and the provision of countertrade, local labour usage agreements, and cheap access to materials from the contractor's home country. While the reasons for these patterns have already been discussed in relation to Table 5.5, one factor from Table 5.6 which surprisingly does not seem particularly relevant within clients' demands is high quality of construction. From interviews it seems likely that clients are reluctant to demand quality separately on the basis that this would raise prices above that acceptable to the client.

5.3.2 **Major firm specific O advantages in international construction**

It is suggested here that the empirical results in Tables 5.3-5.6 tend to substantiate the theoretical predictions of Chapter 4 concerning O advantages. Firm specific advantages in international construction are likely to be embodied in three features of the international contractor which form the base for product differentiation:

5.3.2.1 **Reputation and name of the contractor**
It is clear from the tables that where competition is likely to involve contractors offering similar services at similar prices, product differentiation by reputation (which is embodied in the firm's name) will be the major firm specific advantage. From Table 5.6 it seems we can justifiably argue that the client will demand a solid reputation (in addition to a relatively cheap price) as a means of risk minimisation. As such, name of the firm may be the most important firm specific O advantage even though it may be considered only a reflection of all others.

5.3.2.2 Quality of human capital

Because service sector industries sell expertise and knowledge rather than a physical good, the essential element of maintenance of O advantages to the service MNE will be investment in training, which can be regarded as R&D in human capital (Dunning and McQueen, 1979). Investment in training will enable the service sector MNE to maintain product quality (which will be reflected in the competitiveness of the firm's price) and monitor personnel, and thereby protect the reputation and name of the firm. International construction appears no different from any other service sector industry in this respect. While it may be argued that by stressing reputation the contractor is implicitly acknowledging the skill of the firm's personnel, the predominance of knowledge and expertise as competitive advantages of the contractor in Tables 5.3 and 5.4 illustrates the emphasis that contractors place upon the quality of human capital.

To identify how quality of personnel is maintained in international construction, participants in the survey were asked to outline their training schedule for the training of personnel for overseas operations of the firm, both on and off site. Although the survey can be split into two groups, comprising 24% of the survey who said they carried out no specific training, and 76% who undertake in house training, two factors should be mentioned:

1. All of the contractors in the survey placed emphasis on professional qualifications for staff, and encouraged (and sometimes demanded) specialist education. In the case of the firms which did not carry out training, there was a general confidence in further education of this sort to guarantee a standard level of expertise attained, so that even in these companies quality of personnel was regarded as important.

2. Of the 76% who implied that they carried out in house training, 50% stressed that this was mainly based upon experience and on site training and supervision (where applicable), while the remaining 50% said they undertake firm specific training. This generally involved detailed coursework and experience designed to promote and instill a certain firm specific procedure so that personnel could offer a product automatically linked with the name of the firm. In addition, foreign language training and acclimatisation to the local cultural and social environment took place where it was considered necessary (eg Middle East with alcohol

restrictions).

It would appear then that within the industry emphasis is placed upon quality of personnel as a major feature of competition, although the extent of training may depend upon the objectives of the firm to differentiate its product. In house specific training is more likely where the firm relies upon a particular skill as the firm's product to ensure quality which will also be a source of product differentiation. However in general it is argued that experience and expertise (as designated by professional or in-house qualifications) will be the major components of quality of personnel, which will be reflected in the name of the company.

5.3.3 Size of firm

Although size of firm may be considered within the context of country specific advantages since the size of the domestic market is likely to be positively related to average size of firm (as discussed below), firm specific O advantages will generally be influenced by the size of the firm on technical and financial grounds so that the larger construction firm will have an advantage over its smaller competitors.

Technical aspects of size relate to access of the company to highly qualified personnel (as suggested in Chapter 4) which is herein classed as a major firm specific advantage. If we consider that the larger is the contracting firm the greater are the firm specific resources to obtain the highest quality personnel, then clearly size of the firm may generate a significant advantage.

Financial advantages relate to the greater source of capital (internal and external) that are positively related to size. Access to capital may be regarded as an aspect of the greater ease with which a large contractor may obtain finance, generally at cheaper interest rates. A recent survey for example suggested that fully capitalised highly liquid mature contractors will be able to borrow at prime rates in most commercial markets, while all other contractors will generally be restricted to unsecured loans for short periods only, facing interest rates ranging from 0.5% to 6% above prime, with higher compensating balances (Engineering News-Record, 16/9/82). This obviously restricts the smaller companies and gives the larger contractor an advantage that

Table 5.7: Types of Bonds Required within International Construction

Type of Bond	% of Contract Value	Use in International Contracting
Tender/Bid Bond	Normally 1-2% though can be 5%+	To ensure contractor's bid is serious; bond may be called if contractor does not accept contract after bidding, or if contractor withdraws bid before closing date.
Performance Bond	Normally 5-10%	To protect client against unsatisfactory work by contractor during construction, bond may be called where contractor abandons project or client is not satisfied with work during construction.
Advance Payment Bond	Normally 10%	To ensure that client advance payment is utilised for buying specified materials, equipment, etc. Bond may be called where this does not take place.
Retention Bond	Usually 5-10%	To protect client in subsequent period after construction as a guarantee against faulty workmanship, inferior materials etc. Bond may be called where client not satisfied with final good. Although this may be negotiable where specified within the bond agreement.
Total	21-35% of Contract Value	

in international contracting can increase probability of success at the bidding stage:

1. Given that the preparation of a tender may cost anything up to £500,000 in international contracting (and feasibly more in 'jumbo' projects), the larger contractor is not as restricted in the bidding stage and can bid for more projects than the smaller firm. While this may not be successful in each case, the more the contractor bids the higher the probability of success. Additionally in prequalification, the larger contractor stands a better chance of being selected to bid on the basis of financial standing.

2. The provision of bonds within the industry favours the more financially secure contractor in the bidding and construction stages. Connected with 1, it is often the case that the contractor has to provide bonds drawn on the company's financial reserves that may be cashed by the client where a specific commitment concerning the bond is not met by the contractor. There are four major types of bonds in international construction, as illustrated in Table 5.7. If we consider that bonding may equal anything up to 35% of the contract value, and that additionally the contractor will have to insure the firm against unfair calling of bonds, then clearly the size of contract the firm can feasibly bid for is limited by access to financial resources where bonding is required. In this case the larger firm ceteris paribus is at an advantage to the smaller contractor.

3. The larger contractor has the financial means to undertake extensive VI and HI where this will confer benefits upon the firm. If it is additionally considered that the more diversified the firm the greater are the benefits in terms of global diversification strategy of the contractor (cf Chapter 4), then size of firm may be regarded as an indirect contribution to overall efficiency within the firm.

5.4 COUNTRY SPECIFIC ADVANTAGES

If we argue, as suggested within Dunning's eclectic framework, that the generation of firm specific advantages will originate with country specific factors (ie home country characteristics and the interaction of these with the host country market), then clearly the identification of these characteristics form a major component of how competition has emerged within the industry. The importance of country

Figure 5.1: Overseas Awards of Major Contracting Countries, 1980-84

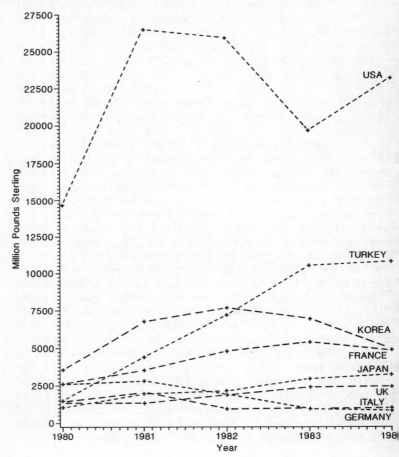

Source: Contracting countries' government departments, cf. Table 6.11

specific factors may be illustrated empirically by comparison of performance of different countries' contractors at the macro level. Given that country specific characteristics influence the effectiveness of that nationality's contractors to compete utilising ownership advantages, where one nationality group performs better

Empirical Analysis of Ownership Advantages

than another then it may be argued that this is a function of the relative usefulness of country specific characteristics for the generation of ownership advantages.

In the context of this argument Figure 5.1, which shows the value of work done by the major contracting countries, illustrates that there are significant differences between the performance of the various nationality groups. These countries were chosen as the major contractors for this empirical analysis because over the period contractors from these countries contributed over 2% of total foreign awards taken by the top 250 international contractors as reported in the annual ENR survey. Any reference forthwith to the major contracting countries refers to the eight countries represented in Figure 5.1.

The most obvious feature of the figure (which is taken from various government publications of the nationality groups) is that American contractors have dominated the industry over the period 1980-1984, despite experiencing fluctuating demand for their services with an overall decline in awards beyond 1981. Turkish contractors have shown an increase which has been experienced to a lesser extent by Japanese and British contractors. Korean and French contractors have maintained a high level of work in the industry, but have faced a declining market in the latter half of the period. Italian and German contractors have experienced a gradual erosion in their overseas awards since 1981.

In terms of the analysis, Figure 5.1 has two implications:

1. The diagram shows that country specific O advantages are likely to be important within the industry. This is illustrated by the fact that there are distinct country specific trends suggesting that some nationality contractors have been competing more successfully in the international construction market than others.

2. Given that some nationalities have increased their overseas earnings while others have seen a decrease in their market share, it is obvious that in a dynamic sense attractiveness of country specific factors may change.

Following from the analysis of Chapter 4 we illustrate here three potential sources of country specific differences which may form the basis of firm specific O advantages.

5.4.1 Domestic Market characteristics

5.4.1.1 Size of domestic market

Figure 5.2 illustrates the value of construction work within the domestic markets of the major contracting countries for the period 1980-1984. From the diagram it is clear that American contractors had the largest domestic market over the period, although the level of construction investment in Japan has been highest overall in terms of per capita and per square metre, as part of extensive urban and rural development in the country (Umeda, 1980). European countries in the group have had fairly stable construction demand, though the magnitude of investment has ranged from Germany and France with moderate to high demand, to Italy and the UK with relatively low demand. Turkish and South Korean contractors have experienced insignificant demand for their services domestically in relation to their overseas activities.

The major benefit of a large domestic market to enterprises from that country is that, ceteris paribus, market size is likely to be positively related to average size of the firm (Dunning, 1981). If this argument is justified in the case of international construction, size of market would provide a significant benefit in the generation of firm specific advantages to contractors from that market in relation to competitors from smaller countries, which may lead to the generation of further firm specific advantages attributable to size of the firm (as argued in Chapter 4 and above).

To test this proposition empirically, a simple regression was carried out using average size of the firm according to nationality as the dependent variable, and size of domestic market as the independent variable. Size of the national market was taken from Figure 5.2 for a specific year (1982), and average size of the national firm was taken as a rough proxy from data in the annual ENR survey. The 10 largest international contractors of each major contracting country were picked on the basis of level of foreign awards, and then averaged according to total awards (domestic + foreign) which is used here as the proxy for firm size. The averages were then regressed as the dependent variable. Table 5.8 summarises the means used as a proxy for average size. Although the figure in column 12 of the table is only a rough estimate of average size, it nevertheless serves to show that on average the firm size may be differentiated according to

Empirical Analysis of Ownership Advantages

Figure 5.2: Domestic Awards of Major Contracting Countries, 1980-84

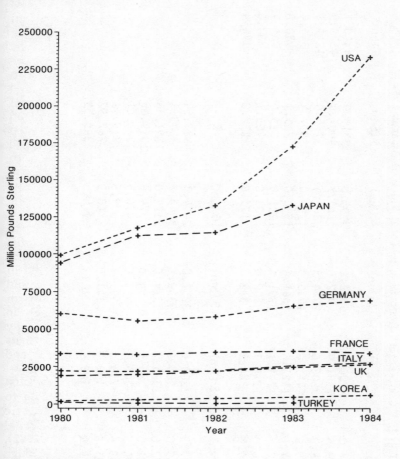

Source: As Figure 5.1.

nationality, with the American contractors being the largest in the survey and the Turks the smallest.

The resulting regression equation based upon the data was:

$$SIZE_{it} = 0.478 + 0.026 \, DOM_{it}$$
$$(1.62) \quad (6.06)$$

$n = 8$
$R^2 = 0.86$

Table 5.8: Size of Top 10* Contractors According to Total Value of Awards 1982

	1 ($m)	2 ($m)	3 ($m)	4 ($m)	5 ($m)	6 ($m)
US	5704.0	8200.0	6507.1	5490.0	3315.0	4319.7
Japan	3069.5	1391.0	982.0	3115.3	4067.0	4379.9
France	1898.0	2808.0	2074.3	2065.0	1714.0	1812.5
UK	3429.6	1451.0	1995.9	1299.0	397.4	345.4
Germany	4864.14	1128.6	1361.0	1990.0	775.7	980.0
Korea	3422.7	1549.1	1201.6	1201.0	915.0	756.4
Italy	3246.0	1914.7	800.0	503.8	409.1	300.0
Turkey	1405.8	384.3	318.3	175.6	232.0	146.4

	7 ($m)	8 ($m)	9 ($m)	10 ($m)	5 ($m) x̄
US	4217.3	3200.0	3252.0	2654.3	4685.94
Japan	469.0	400.3	4352.8	3043.8	3043.8
France	1372.0	1335.7	1314.0	2122.0	1851.55
UK	881.0	95.0	94.0	174.6	1014.09
Germany	370.0	540.9	305.0	490.5	1280.61
Korea	903.6	753.6	911.3	579.8	1219.41
Italy	181.6	123.0	141.0	60.0	767.92
Turkey	114.0	106.4	N/A	N/A	360.35

* NB, Turkey = 8

Source: Engineering News-Record, July 1983

Empirical Analysis of Ownership Advantages

where

$SIZE_{it}$ = Average size of firm in country i at time t
DOM_{it} = Value of construction work in country i at time t

Given that the t statistic upon the independent variable suggests significance, and the r coefficient is reasonably high, we argue here that the simple regression together with Table 5.8 provide justification for the link between size of domestic market and size of average firm, so that market size may be considered a country specific advantage in the context of international construction.

5.4.1.2 Comparative advantage in international construction

Although size of the domestic market may be of importance to the contractors from large DCs, Figure 5.2 suggests that it is not a necessary condition for success in international markets - the Koreans and Turks for example have both been major forces in the international construction industry in the 1980s although neither have a considerable domestic market. The success of these countries, and the failure of others abroad can be put down (at least partially) to comparative advantage of the home country. It has already been suggested in Chapter 4 that comparative advantage may reflect the nature of enterprises within a country, which is also likely to be the case in international construction. If we consider that the international construction industry is a supplier of human capital, then clearly comparative advantage within the industry will be reflected in the nature and skills of contractors according to nationality. In the context of this argument, South Korean contractors possibly provide the best example of exploitation of comparative advantage. Korean contractors have evolved from primarily sub contracting for DC contractors to become one of the major international contracting countries by concentrating particularly on labour intensive general construction in which they can undercut the price of the DC contractors by 'exporting' all the manpower required for overseas projects (including labourers) at extremely low wage rates which reflect the domestic labour environment. Table 5.9 illustrates the effectiveness of this policy and the demand for South Korean services in the Middle East between 1975-1978, when the number of South Korean nationals within the region increased by some 770%. Pakistani, Indian, and more recently Chinese contractors have entered the industry competing in the same environment with a similar advantage.

Table 5.9: Number of South Korean Nationals Working in the Middle East 1975-78

Year	Saudi Arabia	Iran	Other	Total
1975	3000	2400	1200	6600
1976	15900	1500	4200	21600
1977	31800	6600	14100	52500
1978	38100	6900	12600	57600

% Change 75 - 78 = 770%

Source: World Construction, June 1979

Table 5.10: Top 25 International General Building Contractors, 1981

Country	No of Firms	Value of Contract ($m)	% of Total
South Korea	8	4526.3	29.8
France	3	2841.7	18.7
West Germany	1	2468.6	16.2
U.S.A.	3	1263.9	8.3
Japan	2	807.9	5.3
U.K.	2	762.5	5.0
Other	6	2541.4	16.7
Total	25	15212.3	100.0

Source: Engineering News Record, July 1982

Table 5.10, which is an indication of the top 25 international general building contractors in 1981, illustrates both the competitiveness of the Koreans within low technology labour intensive building (which includes general building, highways, airports, marine and manufacturing plant construction) in developing regions, and

Empirical Analysis of Ownership Advantages

also the fact that many of the DC contractors have shifted away from this type of construction because they have not been able to realistically bid against the LDC and NIC contractors in this section of the industry.

Table 5.11: Top 25 International Contractors in Power and Process Plant Construction (1), 1981

Country	No of Firms	Value of Contract ($m)	% of Total
U.S.A.	14	21639.8	73.8
Italy	3	3424.5	11.7
U.K.	1	1370.2	4.7
France	1	800.0	2.7
West Germany	1	361.4	1.2
South Korea	1	210.4	0.7
Other	4	1501.7	5.2
Total	25	29308.0	100.0

(1) Includes chemical process, power plant, and offshore platforms, etc. for contractors with more than 50% of all contracts in these markets.

Source: Ibid.

As a contrast to the Korean case, comparative advantage in the highly capital intensive countries of Western Europe and America is embodied in specialised high technology (high-tech) construction, or construction which requires complex, capital intensive and sophisticated activities within the construction process. Much of the expertise required for this has come from the home country, so that DC contractors may specialise in some particular form of high-tech construction which they have evolved in relation to home market demands. France, for example is considered a leader in nuclear power plant construction, while the Americans have a significant advantage overall in the power and process plant construction industry. US firms have a competitive edge in this field because the construction of power, chemical, industrial, and particularly petrochemical plant is dependent on high technology which is both generated and demanded in the US domestic market.

Table 5.12: Diversity of Services Offered by Top 15* Contractors from Major Contracting Countries, 1982

	Total	Building	Highway/Bridge	Water/Sewer	Manuf./Process	Power
Korea	15	15	15	15	14	13
Japan	15	11	11	13	15	12
US	15	10	8	10	15	11
France	15	10	10	8	11	10
UK	14	11	10	11	9	9
Germany	15	8	8	11	11	9
Italy	15	6	7	8	9	9
Turkey	9	8	4	7	6	5

	Marine	Airport	Design	Constn. Management	$\bar{x}$
Korea	13	15	14	1	12.78
Japan	11	10	13	3	11.00
US	11	8	14	11	10.89
France	10	11	15	6	10.11
UK	11	11	11	3	9.56
Germany	7	6	11	1	8.00
Italy	8	6	13	1	7.44
Turkey	3	2	6	1	4.67

(*NB: UK = 14, Turkey = 9)

Source: Engineering News-Record, July 1983

Table 5.11, which shows the top 25 international contractors in high technology power and process plant construction in 1981 illustrates the effectiveness of this competitive advantage to the US contractors, who took approximately 74% of total foreign awards of the contractors in the table.

To illustrate the effect of the domestic market upon the diversity of services offered by contractors, the top 15 international contractors in each of the major contracting countries were catalogued according to the services they specialise in in international markets, and then averaged, to give the results in Table 5.12. While the averages are interesting in that there appear to be differences between nationalities in terms of services offered, what is of more interest here is the patterns of these services according to comparative advantage. The Koreans for example are generally highly diversified in the services they offer which is attributable to a conscious policy on the part of Korean contractors and the government to provide demand for Korean contractors' services over a wide variety of construction skills, although it is significant that only one Korean contractor in the table offered construction management as a service. This may be a consequence of the fact that because construction management relies upon management skill, Korean contractors cannot exploit their major comparative advantage (ie cheap labour supplies) by offering this service and therefore may not be as efficient as DC contractors in this respect, who are likely to have a high level of management skill. American contractors in the table, for example, were the largest nationality group to offer construction management as a service, which may be considered a consequence of the fact that construction management was first developed in the American market, and also because of the high level of human expertise which is needed for construction management.

Japanese and American contractors in the table were particularly involved in manufacturing and process work, and also in power and design which all require complex high-tech construction skills which reflect characteristics of the domestic market environment. French contractors in Table 5.12 were strongly represented in the area of design, which again may be put down to a national characteristic; the development of specialist in house design work which bypasses the traditional role of the consulting engineer was first developed by the French for use both at home and abroad. The Italians have developed a similar system. The

German and British contractors show reasonable diversity in the table, but no significant patterns in any specialist area, while the Turks do not seem highly diversified, and have not entered the construction management market for similar reasons to those of the Koreans.

From the discussion above it is argued here that in terms of comparative advantage there are three potential sources of differentiation especially between DC and LDC countries which may be attributable to national patterns:

1. Technological knowledge. DCs have a comparative advantage in high-tech or capital intensive construction in which they have expertise that may be utilised in overseas construction. LDC contractors on the other hand will tend to concentrate upon labour intensive simple projects on which they may exploit their advantage of an extensive cheap labour sector at home.

2. Management. Through expertise and training DC contractors are more likely to have the capability and resources for large scale operations requiring extensive global coordination and efficient on site management which have been built up over a number of years.

3. Marketing. DC contractors in general are able to provide a more diverse array of products and services, while LDCs specialise in specific types of projects where low price is the key competitive element. However, as Table 5.12 illustrates, the South Koreans as NIC contractors may provide the exception to this situation.

All three of these characteristics may be considered a macro reflection of the analysis concerning the product cycle in international construction as discussed in the previous section of this chapter, and they are likely to have evolved over time in the developed markets in a learning curve effect. As such, they may form the basis of firm specific differentiation particularly between the DC and the LDC contractor in the international environment.

A further point that is likely to be linked to comparative advantage of contractors is the relationship between the construction industry and other aspects of the domestic market which may prove advantageous in overseas markets. Generally these will fall into three categories:

1. Manufacturers of goods relating to construction Related to comparative advantage, the greater is the demand for home country materials and goods in foreign markets that require some aspect of construction, then the greater is the possibility of a contractor of the same

nationality getting the contract. Japanese contractors for example may have an inherent advantage in the capital goods and machine tool industries of Japan as they are market leaders in their respective fields.

The Japanese and Korean contractors especially may be at an advantage in this respect because in general the large international contractors from these countries are part of much larger trading houses. Japanese contractors such as Mitsubishi, Mitsui, and Kawasaki form part of the massive trading houses known collectively as the 'Sogo-Shosha' in Japan, while the Korean equivalent, the 'Chaebol' includes Hyundai, Daelim, Daewoo, and others that all have extensive construction interests overseas. (2) These construction companies benefit both in terms of having the financial backing of the company as a whole and also from technical linkages enabling the firms to offer a complete package of construction including materials at competitive prices, construction, plant machinery, installation, and after sales service. In the Middle East for example, in the past Hyundai construction has used Hyundai Heavy Industry's experience and capacity in the assembly of steel structures, and also for the provision of cheap materials for construction.

While the Chaebol and Sogo-Shosha provide extensive internal benefits of conglomerate diversification in the provision of in house materials and experience which is not seen to the same extent in other major contracting countries, the provision of specialist materials and equipment within the domestic market is likely to be beneficial to all contractors of that nationality group in the international environment for reasons already discussed, so that the state of a country's manufacturing industry may be reflected by the demand for a country's contractors.

2. Home country clients. Producers who set up production facilities abroad are more likely to demand a home country contractor for the construction of those facilities than a foreign contractor on the basis of first hand experience of home country contractors and links that have been developed through the domestic environment (though this clearly depends upon the supply of necessary skills in the home market). The consequence of this is that where a country's producers are heavily involved in international production, contractors from that country may benefit directly by getting a lead into those markets by constructing production facilities for home country clients. American

contractors for example have benefited in Saudi Arabia by undertaking petrochemical plant and drilling facility construction for ARAMCO, (3) so that within Saudi Arabia only certain US contractors are invited to bid on some types of high-tech oil related projects.

The significance of home country clients to overseas construction was included in the survey by asking participants to outline their initial motivations for entering overseas markets. 40% of the total said that they had entered specific markets to undertake work for home country clients, which suggests that this may be added as a country specific advantage in cases where the contractor originates from a country with a large manufacturing sector geared to international production.

3. Services related to construction. Often services connected with construction will be incorporated within the contractor's product in a bid, so that the greater is the quality of services which may be used by the contractor within that country, the greater are the chances for contractors of that nationality to differentiate their product according to these services. One area in which this is likely to be of particular relevance is in the provision of financial services. Two aspects of finance that have become increasingly important in international construction over the past 10 years are the provision of project finance by the contractor, and countertrade. The consequence of this is that in many cases the contractor will be competing on the attractiveness of the financial package offered, so that banks and contractors have to coordinate their activities more than ever before. This suggests that contractors from countries with large and progressive banking sectors are likely to offer more attractive financial packages, and therefore have a country specific advantage which may directly affect their competitiveness.

Table 5.13 illustrates that UK and US contractors may particularly benefit from an extensive international banking sector, as illustrated by the competitiveness of these countries' banks in the Eurocurrency syndicated loan market. Out of the top 20 lead managers in syndicated loans in 1981, 6 were American and 5 were British implying market leadership in that region of finance which is potentially the source of private capital for project finance.

French and Japanese contractors may benefit from the close coordination that exists between the financial and construction sectors in these countries. Many of the French

Empirical Analysis of Ownership Advantages

Table 5.13: Top 20 Lead Managers in Syndicated Loans 1981

Rank*	Lead Managers in Loan	No of* Loans	Amount (US $mil)	Nationality
1	Bank of America	137	9,794	U.S.
2	Citicorp	151	7,316	U.S.
3	Chase Manhattan	132	5,059	U.S.
4	Bank of Montreal	64	4,223	Canada
5	Lloyds Bank	109	3,371	U.K.
6	Midland/Crocker National Bank	159	3,235	U.K.
7	Royal Bank of Canada/ Orion Royal	103	3,119	Canada
8	Manufactures Hanover	113	2,884	U.S.
9	National Westminster	98	2,775	U.K.
10	Morgan Guaranty	70	2,544	U.S.
11	Barclays	77	2,441	U.K.
12	Bank of Nova Scotia	79	2,389	Canada
13	Bank of Tokyo	102	2,320	Japan
14	Credit Suisse First Boston	21	2,002	Switzerland
15	Continental Illinois	54	1,906	U.S.
16	Credit Lyonnaise	75	1,841	France
17	Industrial Bank of Japan	50	1,661	Japan
18	Banque Nationale de Paris	53	1,595	France
19	Schroder Wagg	12	1,564	U.K.
20	Arab Banking Corp	68	1,550	Middle East

* excludes U.S. 'Jumbo' loans

Source: Euromoney, March, 1982, Dun and Bradstreet, 'Who owns Whom', 1985.

banks for example have substantial share holdings in the major French contracting companies and therefore liaise closely with the demands of the construction sector domestically and overseas, while the Japanese government is involved with the coordination of the Japanese banking sector with all aspects of Japanese industry to promote exports. If we consider that the UK and US banking sectors are not restricted to use solely by UK and US contractors, this theoretically suggests that the Japanese and French contractors have an inherent advantage in this respect over their UK and US competitors, though in practice it seems that both British and American contractors benefit from having a large banking sector. 70% of the survey suggested on this issue that their home country financial institutions provided satisfactory or good assistance to their overseas operations.

5.4.2 Nationality of consultant

When questioned upon the influence of the nationality of the consultant engineer in the success of the firm in a bid, 65% of the respondents suggested that it improved chances of winning a bid if the consultant was of the same nationality as the contractor, while the remaining 35% said that it had no effect or could possibly be detrimental to the firm in the bid situation. In practice, the effect is likely to differ according to the nationality of the consultant. Survey participants were asked to list the designs and specifications that provided most and least difficulty according to nationality of the consultant, to see if this formed a barrier to competition in the bid situation. The least difficulty designs and specifications were those of British consultants (by 58% of the respondents) and American consultants (37% of the total) though given the nationality spread of the survey these results were to be predicted. The designs and specifications most often cited as giving difficulty for compliance were the French (by 47% of the survey), and various others that were not classed as significant to the survey as a whole such as German (16%), Italian (10%), Japanese (10%), Middle Eastern (10%) and American (10%).

The difficulty associated with French specifications tends to reflect the attitude of the French consultants in overseas work; from interviews with contractors and various consultants on this subject it appears that French consultants form part of a coordinated policy of overseas expansion of the French government, and so in general will specify French materials and specifications to be used within construction, and appoint French contractors for construction on the basis that only they can fully understand the consultants' methods and designs, and often language, from first hand knowledge. Other nationality consultants in

Empirical Analysis of Ownership Advantages

Figure 5.3: UK Consultants' Cost of Work in Hand Overseas, 1974-1984

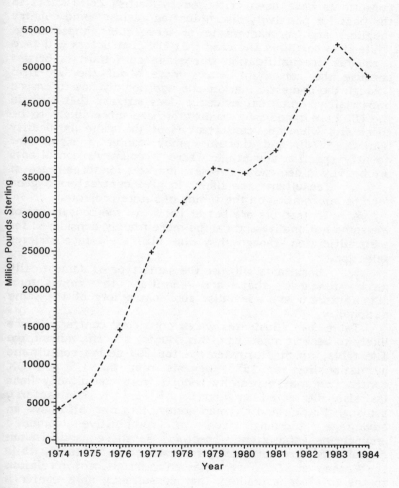

Source: Association of Consulting Engineers

general do not appear to follow such a policy. In the case of UK consultants for example, Figure 5.3 illustrates that over the period 1974-1984 they have been successful in overseas markets, increasing total awards by over 1000%. UK consultants argue however that this success is attributable to the reputation of impartiality, and to positively discriminate in favour of UK contractors would lose them this advantage (Financial Times, 5/3/84). In fact UK consultants have been criticised by British contractors in the past for positively discriminating against home country suppliers and contractors to preserve their impartiality. While it is certainly the case that UK contractors will face a tougher prequalification where the consultant is British because the consultant knows more about the UK firm through the domestic market and additionally has to ensure impartiality, empirical evidence does suggest that even in the UK case tenders by contractors are more likely to be successful when the consultant is of the same nationality (NEDO, 1978), and therefore may confer a significant country specific advantage where a country's consultants are heavily in demand in overseas markets for three reasons:

1. Consultants are likely to give contractors a good feed of information on the timing of future projects.
2. Contractors are better known to the consultants of the same nationality and can be more readily considered for prequalification, though they may ultimately face stricter selection.
3. Consultants will use the same type of domestically used standards that are familiar to contractors, manufacturers and specialist sub contractors of the same nationality.

Table 5.14 illustrates which countries' contractors are likely to benefit most from this country specific advantage. The table, which illustrates the top 200 design consultants by nationality for 1983 suggests that the US and UK contractors may especially benefit from consultancy links (as also illustrated in Figure 5.3 for UK contractors), although French and German contractors may also have an advantage accruing from a competitive domestic engineering consultancy sector. Japanese and Italian consultants were not heavily represented in the table, while the Korean and Turks were classed as insigificant in relation to the total. It is unlikely that consultancy links confer a major benefit on any of these nationality groups.

A further advantage that US contractors may benefit from in this respect, that is not shown in the table, is the work of the US Army Corps of Engineers (USACE) which designs and advises on military construction in nations friendly with America, particularly in the Middle East and South East Asia. While the USACE does offer out many

Table 5.14: Top 200 International Design Consultants in Developing Countries by Nationality, 1983

No. of Firms	Nationality	Foreign Awards $ mil	%
66	American	1204.3	31.3
26	British	592.1	15.4
17	French	360.9	9.4
19	German	253.1	6.6
7	Japanese	127.1	3.3
5	Italian	76.0	1.9
60	Other	1236.8	32.1
Total 200		3850.3	100.0

Source: Engineering News-Record, August 1984

projects to open tender to all nationality contractors, the largest and most sensitive projects are only open to certain approved US contractors to maintain quality of construction and ensure aspects of military and commercial secrecy. In the past USACE has provided some US contractors with a great deal of work in overseas regions, though clearly statistics on the operations of USACE are restricted for security reasons. It is interesting to note that the French have formed a similar version of the USACE, the 'Societe Navale Francaise de Formation et de Conseil' which offers similar services as a means of capturing some of USACE's market and also helping defence and construction sales in this respect.

5.4.3 **Home government support**

Possibly the most commonly stressed factor in interviews with participants in the research was the need for home government help to combat the extensive direct help that some nationality contractors receive. This has already been illustrated in Table 5.4, where the greatest perceived disadvantage with respect to other nationality contractors was the relative level of home government support given. When questioned further on the subject, 90% of the respondents considered that they receive inadequate government support relative to competitors.

We distinguish here between indirect and direct government support as discussed in Chapter 4. However, because the provision of government help in the project finance and subsidisation of construction requires detailed analysis and is regarded as one of the most controversial issues in the contemporary industry environment, it is outlined in more detail in Chapter 7.

Indirect government help may take several forms and ultimately will come down to the attitude of the government towards overseas expansion and help for home country enterprises abroad. Within the group of contracting countries analysed here, France, Japan and South Korea have been particularly interventionist to expand overseas exports, and have accordingly developed coordinated strategies between various sectors of industry so that export of goods and services is often 'en bloc'. With relation to construction, all three have initiated links between construction, the provision of related services and the manufacture of goods connected with construction so that they may compete abroad as a single national entity. This enables the home government to provide extensive help, which will not be possible where contractors of the same nationality compete against each other, because this would inevitably lead to a wasteful loss of government resources and possibly to claims of favouritism on the part of the government. Germany, Italy, and Turkey provide less coordinated policies, while UK and US governments currently follow a laissez faire type attitude in this respect.

Indirect government support is generally likely to take the form of links between governments that may prove beneficial to contractors particularly where government to government negotiated construction awards take place (these will usually include some aspect of project finance provided by the home government of the contractor), or a 'knock on' effect where construction may be part of a wider package. Inter government deals involving the purchase of capital equipment or defence sales (both of which may often be provided through government to government deals) have in the past been major 'lead-ins' for contractors particularly in the Middle East where they have been demanded extensively for commercial development and national security reasons. Table 5.15 illustrates both defence and capital goods sales in the Middle East for the first half of 1984, and suggests that France, America, and Japan may have benefited from these links, and to a lesser extent UK,

Empirical Analysis of Ownership Advantages

Table 5.15: Defence and Capital Goods Sales in Middle East by Nationality of Seller 1984 (First half)

	Defence ($m)	Capital Goods
France	4,000.0	85.0
US	683.8	N/A
Japan	*	350.0
UK	125.0	*
Italy	*	112.0
Germany	*	178.0

*Below $50m

Source: MECDA, 1985

Italy and Germany.

Possibly of more overall benefit to contractors are government links that occur for political and/or economic reasons, and which enable the contractor to be treated favourably on host country public contracts in countries where there is a good political relationship between the home and host country of the contractor (this may be related to financial aid, defence or related goods sales or political or economic 'gifts' to the host country). It should be noted however that the argument goes both ways. Between 1979-80 for example an embargo was imposed against UK contractors tendering for federal contracts in Nigeria as a means of putting pressure on the British government not to recognise the transitional government in Rhodesia (Financial Times, 24/8/81), while American contractors have suffered in the Arab Middle East because of American government support for the state of Israel (MECDA, 1985). Nevertheless, where government to government links are favourable there may additionally be positive commercial advantages to contractors. Arguably the greatest of these is the provision of single taxation (where the contractor faces a single taxation rate rather than pay tax in both the host and home country for overseas earnings), which is likely to confer significant benefits on the contractor in terms of overseas earnings thus the contractor can charge a comparatively lower price for the same return than 'double taxed' nationals.

Direct government support, like indirect support, will

Table 5.16: Home Government Legislation of Direct Influence of Contractors' Overseas Operations

	France	W. Germany	Italy
Taxes Corporate Income Abroad	Limited –	Limited –	No +
Taxes Personal Income Abroad	No +	No +	No +
Rebates Value Added Tax (VAT) on exported materials, equipment and services	Yes +	Yes +	Yes +
Enforces Corrupt practices act	No +	No +	No +
Enforces anti-boycott act	No +	No +	No +
Encourages consortia for foreign projects	Yes +	No –	Yes +
Provides cost subsidies, bonuses, tax credit for foreign projects	Yes +	No –	Yes +
Provides credit insurance	Yes +	Yes +	Yes +
Provides political risk insurance	Yes +	Yes +	Yes +
Ties loans to use of national contractors' architects and engineers	Yes +	Yes* +	Yes +
Provides grants and loans to nationals for feasbility studies	Yes +	Yes* +	Yes +
Provides bid and performance bonds and guarantees	Yes +	Yes +	Yes +
Promotes its nationalised or quasi governmental construction firms	Yes +	Yes +	Yes +
Promotes its nationalised or quasi governmental consultancy firm	Yes +	Yes +	Yes +
Provides access to cheap labour supplies	No –	No –	No –
Provides countertrade facilities	Yes +	Limited +	Yes +

Table 5.16: continued

	Japan	S. Korea	U.K.	U.S.
	No +	Limited -	Limited -	Yes -
	No +	No +	No +	Yes -
	Limited -	Yes +	Yes +	Limited -
	Yes -	No +	No +	Yes -
	Yes -	No +	No +	Yes -
	Yes +	Yes +	Limited*+	Limited* +
	Yes +	No -	No -	No -
	Yes +	Yes +	Yes +	Yes +
	Yes +	Yes +	Yes +	Yes +
	Yes +	No -	Yes* +	Yes* +
	Yes +	Limited +	Yes +	Limited +
	Yes +	No -	Yes +	No -
	No -	Yes +	No -	No -
	Yes +	Yes +	No -	Yes +
	No -	Yes +	No -	No -
	Yes +	Yes +	Limited +	No -

*as part of aid financed package

be a reflection of the attitude of the home government towards assistance in overseas markets, so that the greater is the motivation of the government to support domestic industry's overseas interests, the greater direct support will that industry receive. As suggested above, this will be determined by the aims of the government. For example the Korean government has undertaken several government to government bartertrade deals in the Middle East which have been significant in the maintenance of Korean workloads in the region. These deals involved supplying construction services in return for oil, which the Koreans require because of the lack of Korean domestic natural resources. Additionally the Korean government has been instrumental in the generation of the Korean contractors' major comparative advantage in the Middle East (ie cheap labour supplies) by discharging soldiers on national service early if their skills are required by Korean contractors overseas.

Table 5.16 summarises some of the legislation that directly affects international contractors overseas, and may be classed as competitive advantages or disadvantages. Turkey is not included because details on the government legislation are sketchy and inaccurate. Items are either classed as advantages (+) in the table or disadvantages (-) depending upon the legislation. Items marked 'limited' are not significant enough to present a major competitive advantage or disadvantage.

The major feature of the table is that the level of direct help for international contractors varies. Hence while the French and Italian governments provide extensive help for contractors, American government legislation at the opposite extreme has been so restrictive that it may have actually damaged US contractors' overseas activities. However, on this point it should be emphasised that some factors in the table will be more relevant than others. For example the fact that the Korean government has been instrumental in the provision of cheap labour for Korean contractors may outweigh any of the other government induced advantages or disadvantages in the table for Korean contractors. It is therefore useful to split the components of the table under common groupings and outline the significance of each.

    1. Taxation. Contractors from a country with lower taxation than others will generally benefit from being able to put in lower bids to realise the same expected return as higher tax contractors, so that details relating to corporate

Empirical Analysis of Ownership Advantages

and personal taxation of overseas earnings may be significant in overall costs or profits of the contractor. American contractors seem at the greatest disadvantage in this respect, although advantages accruing to corporate tax deferment are limited to all but Italian and Japanese contractors.

2. Project cost subsidy. The rebating of VAT on exported materials and services and the provision of cash subsidies for overseas projects will have a similar effect as taxation incentives, in that they may lower the costs of operation of the contractor and therefore either allow for a greater profit margin or lower price than contractors without these benefits. The provision of access to cheap labour supplies may be considered in the same context. While the South Koreans are the only recipients of cheap labour supplies in the table, the rebate of VAT and cash subsidy provision varies from country to country so that some lose on one policy but gain on another.

3. The corrupt practices act and anti-boycott act is legislation within the home country which aims to prevent bribery for contract procurement by nationals in overseas markets and ensure that nationals do not default on contracts in operation respectively. Where these acts are operative they will be a disadvantage to the contractor, especially the corrupt practices act which may restrict contractors from using a practice that is widely exploited to gain favour in developing market regions in international construction. America and Japan are the only countries to enforce this act, although there are no penalties under Japanese law if the corrupt practices act is not wholly adhered to. At the other extreme to the Americans, the Italian government not only permits the bribery of officials overseas but also accepts the deduction of such payments from taxable corporate income if they can be justified.

4. The encouragement of consortia and provision of grants for feasibility studies may provide a country's contractors with a significant advantage, particularly when backed by home government support. Where this support is likely to include financial help in the form of loans or aid, the finance will often be 'tied' so that all materials and services needed for construction must come from the loaner country. Support for consortia may also take the form of direct payment of operational costs of the venture. The Italian National Institute for Foreign Trade (ICE) for example shares promotional, research and operating costs of

joint export associations, while France and Japan support joint ventures of small and medium sized companies for international projects through long term zero interest loans for overseas expansion. Britain, Germany, and America provide little support in these fields unless the finance is aid linked, although both the German and UK governments provide finance for feasibility studies which gives contractors from those countries a lead in to certain markets.

5. The provision of credit and political risk insurance are aspects of risk reduction of project and export financing which all the governments in the table provide, which suggests that these factors may be fundamental to overseas operations. This is treated in more detail in Chapter 7.

6. The promotion of nationalised or quasi-nationalised construction or consultancy firms is likely to benefit contractors from that country by increasing information, work and lead-ins to certain markets. France, Germany and Italy all have state owned or controlled construction companies, while the Korean government patronises and supports the Korean Overseas Construction Corporation (KOCC) which bids for international projects as a national entity and distributes work among members according to their capabilities and workload. In the context of nationalised consultancies, Britain is the exception since the UK government has no state controlled consultancy for overseas work (NB the US has a state owned consultancy in the USACE).

7. Where bid and performance bonds and guarantees are provided, size of the individual contractor's financial resources becomes less of a constraint in overseas work (cf 'size of firm' section above). Only the Korean and American contractors are restrained by lack of government support in this respect, despite relentless campaigning by both.

8. The ability of the contractor to draw upon government countertrade (or bartertrade) facilities may be particularly instrumental in the success of contractors in regions where payment is in some form of raw material. France, Italy, Japan, and South Korea all have extensive government countertrade facilities which in the past have been used in government to government negotiation of construction and other services in return for raw materials demanded for home country use. Germany and Britain do not deal directly in countertrade but have advisory services that coordinate contractors with experienced commodity

Empirical Analysis of Ownership Advantages

brokers, many of which operate out of the London financial market. American contractors receive no official support in this field.

The usefulness of home government support as a country specific advantage will be determined by the level of support received, so that the greater the government help provided relative to other countries the more this may be regarded as an exploitable country specific advantage in international competition. In the context of this argument this will not only relate to the aspects of support as outlined above, but additionally to financial help to contractors in the form of project finance, which may be a major feature of international competition. For this reason Chapter 7 is devoted to government provision of project finance.

5.4.4 A summary of country specific factors in international construction

Table 5.17 illustrates a summary of the discussion above with relation to country specific factors which may affect the international operations of the major contracting countries. While the table is a self explanatory brief illustration of some of the national factors that may generate ownership advantages, two points should be noted with respect to the table:

1. The source of firm specific O advantages in international construction is country specific factors. Product differentiation and price competition will ultimately come down to factors of the home country that the contractor can exploit, particularly where these differentiate the firm from other nationality contractors. This suggests that contractors of the same nationality are not likely to benefit from competition amongst themselves overseas, since this will reduce the help they receive from home country institutions, particularly government help.

2. The table suggests that country specific factors cannot be isolated from the locational interaction of the home and host country so that the attractiveness of country specific advantages must take account of locational features. The eclectic paradigm is therefore justified in regarding the interaction of ownership and locational factors as a feature of international production, in this case within the construction industry.

Table 5.17: Summary Chart of Major Contractors' Country Specific Advantages

US

Comparative advantage

Extensive highly capitalised market that is at the forefront of many industries associated with technology. The market has large national demand for construction and therefore has the opportunities for extensive training and development of size of firm.

Other industry relations

US contractors may generate substantial demand from US client overseas operations since US industry has an extensive worldwide spread, US industries supporting construction are generally capital intensive that have given US contractors a significant advantage particularly in petrochemical and power construction. Demand for US consultants (and USACE) is high and may provide additional advantages.

Home government support

The US government is either accepted or rejected on political grounds by other countries, but where accepted US contractors may benefit from close political links generated through government to government deals and regulations. Direct support for US contractors is poor in relation to other countries' policies.

UK

Comparative advantage

UK contractors may benefit over LDCs in the fact that Britain is more capital intensive (which is reflected in aspects of construction), but this does not provide a significant advantage in relation to other DC contracting countries who are equally more capital intensive. Market demand is lower than most

European countries as a conscious deflationary policy of the government.

Other industry relations	UK consultants are heavily in demand overseas which may benefit UK contractors. UK manufacturing industry has both highly competitive and uncompetitive states such that generalisation about benefits is difficult. UK financial sector is extensive and competitive and may provide an advantage for UK contractors.
Home government support	Britain has many political and diplomatic links with ex-empire countries, though the provision of economic and political incentives is more limited than that of the US. UK direct policy affecting contractors tends to be 'laissez faire', but services are fairly comprehensive relative to the US though financial help is limited.

FRANCE

Comparative advantage	Apart from the LDC/DC advantage as expressed above for the UK, the French have an advantage in being very co-ordinated between various sectors of the economy, so the construction industry benefits from links with the manufacturing and financial sectors. Market demand is high relative to other European countries.
Other industry relations	French consultants tend to be nationalistic and this tends to sum up French overseas activity so that materials etc are generally all French. France is considered a leader in nuclear power and this has had beneficial effects upon contractors.

Home government support	Very extensive, both directly and indirectly so that French contractors in the past have been likened to an exported nationalised industry. Government to government links are often used to gain construction deals particularly in defence and other sales arrangements. France also generates ex-empire links particularly in Francophone Africa.
GERMANY	
Comparative advantage	Similar situation to UK and France in terms of LDC/DC comparative advantage. Market is fairly large (cf. Figure 6.2) and affluent due to advanced and competitive industrial sector.
Other industry relations	Germany's manufacturing industry and machine tools industry is noted for technical excellence and reliability that is likely to have helped German contractors overseas. German consultants are demanded abroad but relative to US, UK and France take a minor share of the market.
Home government support	Germany still has links with several African states through C18th colonial expansion, and the government is willing in many cases to provide aid linked finance for overseas project. Direct support for German contractors is similar to that of UK contractors and has similar financial resources in this respect.
ITALY	
Comparative advantage	Italy is relatively poorer than the other European contracting countries, though is highly capital intensive in some fields. However, Italy still retains a lower wage level than European competition and therefore maintains labour intensive

	construct is popular in the home market and has been utilised overseas successfully by Italian contractors.
Other industry relations	Chemical plant and steel plant construction are specialities of Italian contractors that may reflect the domestically predominant industrial environment. Italian consultants are not heavily in demand overseas.
Home government support	The Italian government has been instrumental in getting work for contractors in Libya and Nigeria especially through barterdeal and negotiations. The Italian government indirectly supervises the industry through state owned contractors and consultant firms, and while direct support is relatively comprehensive in practice it is limited by the financial resources of the government.

JAPAN
Comparative advantage

	Similar to Germany, Japan has a comparative advantage in manufacturing industry and machine tools that has been advantageous to Japanese contractors, especially since Japanese goods tend to be highly price competitive. Japanese industry is run by the Soga Shosha that suggests that financial and commercial backing is likely to be extensive. Domestic construction demand is extremely high due to need for development in both urban and rural sectors.
Other industry relations	Japan's manufacturing industry allows Japan's contractors to compete in the specialist market of plant installation and construction. Additionally Japanese efficiency at the construction of nuclear power plant is an advantage. Japanese

	consultants are not heavily in demand overseas, (possibly because many of the larger contractors undertake design and building), but generally tend to be nationalistic in a similar vein to the French.
Home government support	Through MITI (Ministry of International Trade and Industry), the government follows export goals that are generally carried out through the Soga Shosha. Japanese construction firms benefit from direct intervention by the government operations.

KOREA

Comparative advantage	Korean contractors benefit especially from having a large cheap indigenous labour sector that has in the past been used extensively in overseas operations to undercut DC contractors especially in the Middle East. Domestic demand is small. Contractors form part of the 'chaebol' which is similar to the Japanese Soga Shosha.
Other industry relations	Through the chaebol the major contractors can obtain cheap materials and services related to construction that helps Koreans maintain a low price. Korean consultants are not heavily in demand overseas.
Home government support	Korean government support is fairly extensive directly and indirectly as part of the government export promotion policy. This includes the provision of cheap labour, and government to government negotiations including countertrade. In practice financial support is limited by lack of foreign currency.

TURKEY

Comparative advantage

Turkey has both a large cheap labour sector and experience of capital intensive construction through past experience, though this is to a lower level than US and Europe. Nevertheless as a result Turkish contractors are generally cheaper when in competition. Additionally Turkey is situated in the Middle East and is Moslem which may have been especially significant in Turkey's share of the Middle East market.

Other industry relations

Turkey has a fairly comprehensive construction material industry which provides cheap materials for overseas construction firms. Industrial links and the use of Turkish consultants are limited.

Home government support

The Turkish government actively supports Turkish contractors and in the past has undertaken extensive training programmes for nationals, several government to government negotiations of contracts in the Middle East via political and diplomatic links, and ensures bidding is carried out as a national entity with full government support. The government however does not support an export insurance scheme and is restrained by limited financial reserves.

Sources: Fieldwork; Atsumi (1982); Engineering News-Record (Sept. 1984); Barna (1983); Nicholson (1982); MECDA (1985); Turkish Daily News (1985); Cezik (1980); Financial Times (May 1985, March 1984, August 1981, September 1980); Civil Engineering (November 1979).

5.5 FOOTNOTES

1. This was necessary because of the limited finance available for the research.
2. In connection with this it should be noted that in Table 5.8 the Koreans show construction awards only, and therefore understate the potential of advantages accruing to size of the trading houses.
3. ARAMCO was initially set up as a joint enterprise between prominent US oil companies and the Saudi government for the exploration and exploitation of oil reserves in Saudi Arabia. Although it has been under full Saudi ownership for 15 years, many of the large US oil companies are still involved with the venture and much of the technical expertise within ARAMCO is provided by American and British expatriates.

Chapter Six

INTERNALISATION AND LOCATIONAL FACTORS

Although from Chapter 5 it is clear that O advantages are of major importance within competition in international construction, the work of Dunning (1981, 1985), Casson (1979, 1982) and Buckley and Casson (1976, 1981) illustrates that these are only part of a wider spectrum of factors that influence enterprises in their involvement in overseas markets. In keeping with the framework of the eclectic paradigm as adopted throughout this research we separate the empirical analysis of these other factors into two groups, internalisation and locational characteristics, which will be interrelated with each other and with O advantages. Results for this empirical study are from the same sources as those discussed in Chapter 5.

6.1 INTERNALISATION FACTORS

Because the name of the firm is the international contractor's most saleable asset since it significantly enhances the firm's chances in the bid situation and is easily transferable between companies (as discussed in Chapter 4 and reasoned from empirical results in Chapter 5), then clearly the choice between exploiting the advantages of the external market and internal hierarchy revolves around the willingness of the firm to sell or lease out this firm specific O advantage. The empirical results from the survey substantiate the theoretical predictions that in international contracting the costs of the external market are too great for externalisation to be considered - none of the contractors questioned had ever licensed out their name for use by others. Table 6.1 gives the justification for this policy for the survey respondents. The major reason cited for not considering licensing, given by 40% of the survey, was that the contractor does not possess a standardised product that can be licensed, although difficulty of ensuring product quality under license and difficulty of writing an effective control contract with the licensee were also common answers (by 25% of the survey in both cases). Fear of underperformance of the licensee, and inability to identify a suitable licensee were each given by 5% of the survey.

Table 6.1: Participants' Reasons for not Undertaking Licensing of the Firm's Name

	%
Lack of standardised product to license	40
Difficult in controlling product quality of licensee	25
Difficult in writing effective control contract with licensee	25
Fear of underperformance of licensee	5
Inability to identify suitable licensee	5

Source: Fieldwork

It is argued here that these results are in line with the theoretical predictions in Chapter 4 since they stress the problems of potential underperformance, quality control, and inability of the contractor to control exploitation of the advantage which are potential costs of the external market. A possible explanation for the low consensus of results in Table 6.1 may be that contractors are unlikely to formulate policies with the same rigorous analysis afforded by internalisation theory, but rather will rely upon subjective valuation. However, the lack of licensing generally in the industry and the results in Table 6.1 suggest that the internal mode of organisation is preferred, and therefore has distinct advantages over the external market. Internalisation theory may therefore be justified in this context.

Internalisation and Locational Factors

6.1.1 **The internal hierarchy in international contracting**

Using the Buckley and Casson (1981) model in Chapter 4, it was predicted that while licensing will not generally be a feasible option in international contracting, exporting and FDI may be utilised either exclusively or simultaneously in any market because of the similarity of the two. Hence in markets where a local presence is required but the contractor wishes to minimise fixed assets, the contractor may maintain a local operation for administrative duties and 'import' on site personnel when required for specific projects. This will not only help maintain low fixed assets in any one country, but also allow for fuller utilisation and exploitation of the firm's personnel by ensuring they are in constant use worldwide (Casson 1985c). Although there is clearly no one system or hierarchy which all enterprises rigorously follow, it was possible to discern common forms of hierarchy from the interviews. For the sake of clarity we define the major components of the internal hierarchy in Table 6.2. The headquarters (HQ) or parent firm will form the ultimate source of resources, finance, and policy that may be passed down through the regional HQ (RHQ), subsidiary, or directly to the project office (PO) in a 'one off' project situation. It should be noted however that these definitions are static and may not fully accommodate the functions of those components over time. For example a local office (LO) that has the capability for autonomous construction up to a certain level of construction (usually specified by the parent in terms of size of contract the LO may carry out) may be regarded as a subsidiary up until this point, although beyond that level the LO may be required more as a local administration base for personnel imported for a specific project. The definitions and characteristics in Table 6.2 are therefore not a rigid framework.

Given the components of Table 6.2, Figures 6.1-6.4 summarise the major relationships within the hierarchy in international contracting. The diagrams, which are a basic chronological chart of the interaction of the various components in Table 6.2 are obviously a simplification of the actual situation but are useful for highlighting the different forms of operation the international contractor may take. These are by no means exhaustive since in reality information, financial and personnel flows are likely to be continuous in all cases so that options are numerous.

Figure 6.1 illustrates the case of 'pure' FDI, where the

Table 6.2: The Hierarchy of Control of the International Construction Firm

Function	Authority	Duties
HQ/Parent Company		
To coordinate and manage the overall policy and objectives if the firm is to follow a global policy that integrates all personnel and actions.	The HQ will have overall authority of all offices and subsidiaries of the firm.	Overall management that will vary according to the relationship with the office or subsidiary in question.
Regional Office		
To coordinate those policies of the HQ with local operations of the firm in a specific area of the world.	Authority over local interests of the firm, though will be responsible to the HQ.	The same as those of the HQ for a specific region only.

Table 6.2: continued

Subsidiary

To work in a certain market as an autonomous branch of the firm with its own personnel and capital source.	Ultimately to the HQ, though will generally act autonomously on all the general directives.	To maintain and improve the firm's presence and work load in a national market.

Local Office

To carry out administrative duties in one country to minimise the work load on the project team.	Will work in contact with the project office and additionally be responsible to the HQ or regional office (where applicable).	Market surveillance and research; all administrative duties necessary within that country to ensure the construction team is unburdened with problems other than technical factors.

Project (on site office)

To carry out construction work.	Responsible to local regional or HQ office depending upon arrangement.	All on site construction.

Source: Ibid.

Internalisation and Locational Factors

Figure 6.1: Subsidiary Operation of the International Contractor

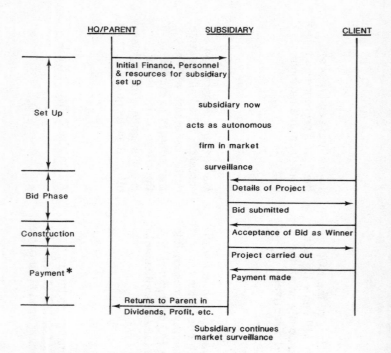

* May take place through construction period

parent initially finances the setting up of a subsidiary which then continues in that market as an autonomous enterprise (as suggested in Table 6.2). The movements between the parent and subsidiary are likely to be capital flows to and from the subsidiary, high level management and firm policy and objectives. Apart from this the subsidiary acts as a separate entity in the procurement and production of construction projects. This form of organisation is likely in 'stable' markets where the parent does not have to coordinate resources as much with other offices and subsidiaries of the firm to maintain efficiency.

In Figure 6.2 the HQ or RHQ (since they have a similar function in this respect) maintain market surveillance, bid

Internalisation and Locational Factors

Figure 6.2: Pure Export or 'One Off' Operation of the International Contractor

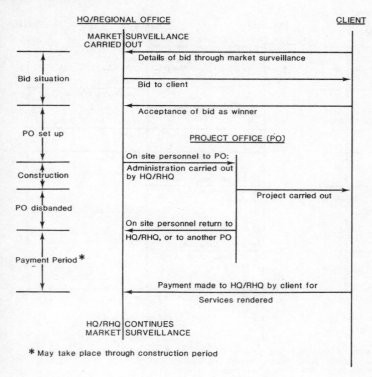

* May take place through construction period

preparation and other administrative duties from outside the market. If a bid is successful in that market the firm exports personnel to form a PO that deals with the construction of the project for the duration of the contract. Once finished, the personnel either move back to the HQ/RHQ or to form or join another PO. Figure 6.2 therefore illustrates what is termed here 'pure' exporting (1) since the production process is only in that market for the period of construction. This policy may be particularly effective in unstable markets (ie spasmodic demand or high risk conditions), since the firm's assets remain fairly mobile throughout construction and are then removed from the market, although clearly this policy limits the possibility of extensive market surveillance and the advantages of

Figure 6.3: Export and F.D.I. Situation Where Local Office has Construction Capabilities (I.E. LO/PO)

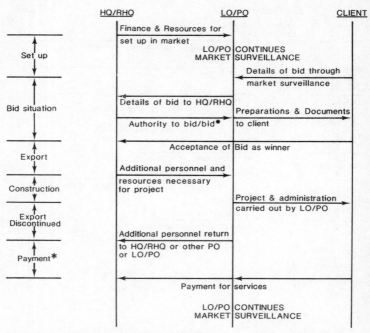

* i.e. may come from either HQ/RHQ or LO/PO depending upon arrangement
* may take place through construction period

maintaining a local presence.

Figure 6.3 is a combination of Figures 6.1 and 6.2 where the LO has PO capabilities and therefore in certain circumstances may be a subsidiary operation. However, where the skills or requirements demand greater resources than the LO/PO possesses, generally the bid will be passed on to the HQ/RHQ for either confirmation of bid or tender preparation using more detailed and extensive resources. If the bid is successful this will lead to the simultaneous exporting and FDI situation (as introduced in Chapter 4) where the LO/PO personnel accommodate and work with additional personnel imported for construction of that project, while maintaining their own duties (especially the LO administration duties to minimise problems for the PO

Figure 6.4: Export/F.D.I. Situation with Separate LO and PO Facilities

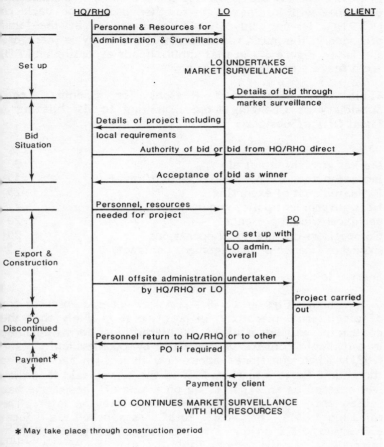

* May take place through construction period

during construction). Once the project is finished the imported personnel may move onto a new project elsewhere, and the local LO/PO may undertake similar duties as before until the next large contract is won.

In Figure 6.4 the situation is similar to that in Figure 6.3 but the local presence is an LO only for administrative and market surveillance purposes. In this case if a bid is considered feasible by the HQ/RHQ and is then successful, a specific PO will be set up for the duration of the project which will benefit from the administrative facilities of the LO. Once the PO has completed the project the construction personnel will be moved to another project or back to HQ/RHQ, leaving only the LO to carry on as the local presence in the market. This form of hierarchy is therefore similar to Figure 6.3 in that simultaneous exporting and FDI are a feature of the organisation.

Table 6.3: Participants' Reasons for Opening Local Subsidiary or Changing from Exporting to Establishing a Local Office/Subsidiary

	%
Market growth/potential	90
Increased need for personal presence/image	55
Increased cost of exporting	25
Host government policy requiring permanent local presence	20
Competitors opening local operations	20
Expiry of licence or management contract	5

Source: Ibid.

Although Figures 6.1-6.4 are only general observations of the internal hierarchy, in practice it is likely that the contractor will adopt all of the hierarchical forms outlined in a dynamic context as suggested by Buckley and Casson (1981). When questioned about how their firm had entered the Middle Eastern market, 95% of the survey said that they had entered as an exporter (ie on a 'one off' basis) and had subsequently set up a local office/subsidiary later. The remaining 5% of the total established a local office/subsidiary on entry to the market without an initial introduction to the region. Table 6.3 summarises the major factors suggested as the motivation for undertaking subsidiary operations or changing from exporting to the establishment of a local presence as given by the survey respondents. Obviously the most important factor according to the table is market growth or potential given by 90% of the total (which is the major determinant in the Buckley and Casson model), followed by the need for a personal presence or image in the market by 55% of the survey reflecting the nationalistic bias of local clients as discussed in Chapter 4. Increasing costs of exporting, competitors opening local operations, and host government policy requiring a local presence were also mentioned as incentives for setting up a permanent local presence, although the significance of these overall is limited. Only 5% of the total entered the market on expiry of licence or management contract, though it is likely that management contract rather than licence expiry provided the impetus given the form of internalisation in the industry.

To analyse if contractors in the survey have organised their internal hierarchy in a similar way to that outlined above and predicted by the Buckley and Casson framework, participants were asked for details of the relationship between the HQ or parent company and the contractor's Middle Eastern offices or subsidiaries with respect to bidding procedure and employment, since these factors are thought to summarise the major links between the components of the internal hierarchy. The results are presented in Table 6.4. The table illustrates that there is no single policy used by the participants on bidding procedure, although the majority of the survey did use HQ/subsidiary information links to coordinate bidding, only 10% of the survey have totally autonomous subsidiaries in the region. In the context of the definitions in Table 6.2 the distinction between subsidiary and office may therefore be negligible for many of the contractors in the survey since it appears that for these companies the local representative undertakes both local construction (which is consistent with subsidiary involvement) and administration and market surveillance for the HQ (which we regard here as a LO) and therefore embraces both functions. This is reflected in the second half of the table which shows that the majority of the participants do operate a LO/PO type arrangement with autonomy up to a 'ceiling' as defined by the parent, beyond which the representative may act as a LO for imported personnel (ie Figures 6.3 and 6.4). It should be noted, however, that participants with only a LO in the Middle East also seem fairly common, possibly reflecting the need for a local presence for market surveillance given the spasmodic demand conditions the contractor is likely to face in the market.

From the survey results we can therefore argue that there are two features of international construction that are of relevance to discussion of the internal hierarchy in this context:

1. The differences between a subsidiary and LO in many cases are likely to come down to the function of that

Table 6.4: Relationship Between HQ and Middle Eastern Subsidiary of Participants in Survey

(a) Bidding	%
Subsidiary bids for local contracts to be checked by HQ	40
HQ bids on subsidiary information	20
Subsidiary has autonomy in bidding up to a 'ceiling' beyond which HQ takes over	20
Subsidiary bids for local contracts on HQ instructions	10
Subsidiary operates as separate entity altogether in bidding	10

(b) Employment |
---|---
Subsidiary employs permanent personnel plus additional HQ | 55
Personnel (or from another subsidiary) at times of high or specialist demand | 0
Subsidiary employs nucleus staff with imported site and other personnel when required for construction | 40
Subsidiary employs permanent home country personnel only for all construction | 5
Subsidiary employs own personnel for all projects | 0

Source: Ibid.

office at a specific time. In practice in international construction the local presence of the firm may alternate between carrying out subsidiary and LO roles, which will be determined by the parent company's regulations and policies and the market environment (ie location characteristics). International contractors therefore have a fairly flexible internal organisation that has great scope for coordination of information and functions between regions and countries to provide efficient utilisation of resources. In terms of the Dunning (1985) model, the use of the internal organisation for the international contractor therefore provides transactional advantages associated with the ability of the firm to capture economies of locational market diversification.

2. The Buckley and Casson model seems empirically justified within international contracting, not only in a

dynamic sense but also because the use of simultaneous exporting and FDI is common in the industry. This is substantiated by the fact that 95% of the contractors in the survey undertake this policy in the Middle East to some degree. This provides further justification for the use of a flexible system by the contractor.

6.1.2 Non equity involvement in international construction

When applied to the international construction industry, one of the major predictions of internalisation theory is that while the contractor is not free to externalise the firm's major O advantage (ie the firm's name) without fear of damage, the characteristics of this advantage are such that it may be protected from fraud, theft or imitation by law. The theory therefore suggests that certain forms of non equity participation may be acceptable since they minimise risks of the external market but maximise benefits. Joint venture and management contract are the most common forms of non equity participation in international construction, and are possible because they allow the firm to retain product quality supervision throughout construction (which we have argued is central to the internalisation issue), but also enable the enterprise to benefit from the lessening of risk exposure, pooling of financial resources, reputation, and skills with other enterprises which will provide the initial impetus to the venture. We restrict the analysis of non equity participation here to that of the joint venture in international construction, although the reasoning behind the adoption of management contracting in the industry follows a similar argument.

Empirical work on the extent of joint venture in international construction tends to suggest that the theoretical prediction that joint venture will be common in the industry is valid. In an ENR survey in 1981 (30/4/81) out of 142 top international contractors and 80 major US contractors, 98% and 91% respectively said that there would be more joint venturing in the future, while a survey in the previous year along similar lines found that 80% of the foreign and 75% of the American contractors would be willing to be included in more international ventures. In the survey carried out for this research the results were overwhelmingly in favour of joint venture. All of the contractors in the survey said that they would consider joint venture on a project, and the majority of these had already been involved with joint ventures several times in various markets.

To analyse the motivation behind joint venturing in international contracting, participants in the survey were asked to list what they considered to be the major advantages and disadvantages associated with such a venture. Table 6.5 illustrates the answers given by the respondents. In line with the analysis of Chapter 5 which suggested that nationalistic bias may be prevalent in the Middle East, Table 6.5 illustrates that possibly the major advantage of joint venture for a contractor entering a foreign market for the first time is an introduction to the business contacts and systems of that market by a locally based partner, which may reduce the bias against that contractor and also increase the effectiveness of market surveillance. Although in some regions this alone may provide reason for joint venture with a locally based contractor, in the Middle East legislation demands that the foreign contractor take an indigenous partner in order to operate in that market, and while there are certain disadvantages to this (as discussed below) 65% of the survey said that this type of arrangement is advantageous (though advantages were mainly associated with the ability of locals to get contracts rather than the combination of construction skills of the partners). Other major advantages of joint venturing cited revolve around the benefits to the contractor of combining resources with another contractor (ie diversified risk, increased reputation, and access to skilled personnel not available within the company). In the table, home government policy was not classed as a major advantage because for the majority of the contractors the home government provides no support or does not have an integrated policy on the use of joint venture by national enterprises. The fact that advantages attributed to a combination of plant holding and financial resources were only perceived as important by 10% of the survey was an unexpected result considering the nature and impetus of joint venturing, although this result may have occurred because the other benefits mentioned may encompass these advantages.

The most common disadvantage cited in the survey was that of fear of underperformance by partners in joint venture. In the case of underperformance by a joint venture

Internalisation and Locational Factors

Table 6.5: Advantages and Disadvantages of Joint Venturing

	(a) Advantages of Joint Venture	%
	Introduction to region by local/known contractor	70
	Local government policy	65
	Diversified risk in politically unstable areas	60
	Increased joint reputation when bidding	55
	Need for specialist skills not available within the company	35
	Home government policy	15
	Plant holding, financial strength and prequalification chances increased	10
(b)	Disadvantages of Joint Venture	
	Fear of underperformance of partners	70
	Local government policy	40
	Inability to find suitable partner	35
	Lack of co-operation by partners in the past	10
	Incompatibility of partners	15
	Home government policy	5
	Unlimited responsibility for partner	5
	No incentive from past experience to joint venture	5

Source: Ibid.

partner, the other partner will suffer since the failure will be linked to the whole venture rather than the firm responsible, so that a contractor in joint venture may face problems of underperformance which involve similar risks to those associated with the licensing of the firm's name. However, the contractor will have a greater overall level of control and supervision in joint venture, and may choose a partner according to past expertise (although this may not be easy, as indicated by the fact that 35% of the total suggested that inability to find a suitable partner was a major disadvantage), so that it can be argued that joint venture does not expose the contractor to the same element of risk as licensing.

The fact that host country government policy is classed both as a major advantage and disadvantage in the table suggests that there are elements of each within the legislation in the Middle East concerning the requirement of an indigenous partner for foreign contractors. While it is true that a local partner will increase the probability of success in bidding with local private and public contacts and a knowledge of the local business system, many contractors are against this policy because often the local partner will demand a large share of the profit for minimal input. Additionally, where the indigenous partner does get involved in the actual construction (which is not always the case) the level of expertise is frequently low so that the major contractor will have to maintain strict quality control to prevent underperformance.

Major problems of joint venturing revolve around the fact that in many cases there are problems of coordination between the venture partners so that efficiency of the venture may be impaired. This is reflected throughout the disadvantages cited in Table 6.5 (including the possibility of underperformance of a partner as given by 70% of the survey), although only 15% directly pointed out that incompatibility may present a problem. While this is a factor that may affect the internal benefits of a joint venture, choice of a suitable partner, careful planning, and continual coordination of the venture's objectives, policy, and management may minimise this potential diseconomy of joint venturing. A more serious problem that is not mentioned in the survey but is relevant to internalisation theory is that in certain cases a joint venture may actually be harmful to the preservation of the firm's O advantages. It has already been argued in Chapter 4 that the skills and expertise possessed by the contractor form the basis of firm specific O advantages, since these are instrumental in the success and reputation of the firm. Because expertise is manifest in human skill and experience, there will be a 'learning curve' effect, such that the contractor with that expertise may face the threat of dissipation of knowledge to a joint venture partner with lesser skills in that field to the extent that the partner becomes a future competitor. This is likely to prevail where the lesser skilled contractor (usually a LDC or NIC contractor) is in joint venture with a contractor with expertise in high tech, or human capital intensive fields. In this case the lesser skilled contractor may by imitation and experience learn to carry out the more sophisticated construction demanded so that the high tech contractor is no longer necessary as an input on future high tech construction projects. The South East Asian contractors, who may receive home government help or backing are noted for following this type of policy (though the Middle East legislation as discussed above is a similar attempt at technology transfer). Westphal et al. (1984) for example notes that in the South Korean case:

> (a) force motivating some joint venture investments has been the desire to gain access to foreign technology through collaboration with foreign companies ... (this has) been a factor underlying the formation of joint venture construction firms in conjunction with contractors from developed countries. (Westphal et al. 1984, p.37).

In terms of internalisation theory this clearly places a cost upon the DC contractor in a joint venture, which must be taken into consideration when deciding upon the form of involvement to take. The theoretical framework suggests that three options are possible:

1. Ensure that technology transfer is limited in joint venture by various measures and policies designed to protect specialist knowledge.
2. Do not enter a joint venture where this type of danger is present.
3. Internalise the potential partner's share of the joint venture to minimise external costs and risks.

While the first two strategies may not be a practical solution for the contractor wishing to maintain workload in overseas markets in the long run, the third policy implies the use of horizontally integrated (HI) strategy to internalise those aspects of construction that are required by the firm but cannot be externalised without exposing the contractor to some form of risk as discussed above. The use of HI within the industry has already been illustrated in Table 5.12, which shows that the diversification of contractors into various specialist construction skills is fairly wide though the degree of this varies with the nationality of the contractor. While Casson (1985c) puts this down to the internal benefits the contractor may gain from HI in the form of full utilisation of assets, possibly a more valid argument for integration of the various components of production is that put forward by Buckley and Enderwick (1985). Using a similar argument to Casson's, Buckley and Enderwick suggest that the firm should only internalise those factors that provide a means to 'the development and maintenance of comparative advantage in tasks' (Buckley and Enderwick 1985, pp.17-18).

Since comparative advantage will reflect the level of development of the contractor in relation to other nationalities (which in effect is the dynamic framework of the product cycle as discussed in Chapter 5), internalisation and joint venture policy will reflect the nature of O advantages of the firm. Korean contractors for example will internalise their labour force since the practice minimises risk of supply of cheap labour, while DC contractors will internalise only their managerial staff which form the human capital element of the contractor's product. Accordingly, HI strategies (that may aim either to exploit or minimise joint venture) may differ subject to the internal benefits that this will confer upon the contractor (ie transactional advantages in the eclectic framework). The extent of these benefits is likely to be determined by two possible factors:

1. The nature of the O advantages. The more country specific the advantages of the contractor, the more incentive there is to internalise that feature to compete overseas with other nationality contractors who do not possess the advantage. Internalisation therefore may follow national patterns according to the nature of the C advantages contractors from that country possess.

2. The degree of multinationality. The greater is the multinational diversity of the firm, the more likely is internalisation of O advantages to fully exploit the transactional benefits of those advantages in the internal organisation. In practice DC contractors are likely to have a greater degree of multinationality than LDC contractors because they have more experience of overseas activities and markets through past knowledge of international construction.

It is likely that both factors may affect the decision to internalise, so that country specific patterns of internalisation may be present in the international construction industry, leading to different internal strategies and objectives. The Korean contractor for example is likely to aim for overall diversity into several sub markets to maintain full utilisation of labour, while at the same time aim for technology transfer from DC contractors through joint venture as a means of maintaining workload in an increasingly competitive market and allowing the Korean contractor to enter more geographically diverse markets to exploit advantages of multinationality. DC contractors on the other hand will aim to incorporate more high tech skills in their product above those possessed by LDC and NIC contractors because their advantage lies in expertise rather than price. Additionally these contractors are likely to have a more efficient internal organisation because of an experience factor that is not available to LDC contractors, so that the DC contractor's decision to internalise the management team will reflect the need of the firm to retain its human capital. Internalisation theory therefore suggests that joint venture for the DC contractor will be most advantageous where skill and expertise of the contractor is increased by the venture, since joint venture with a lesser cost LDC contractor may have short run benefits but is likely to have longer run costs from the erosion of the DC contractor's major advantage.

The Buckley and Enderwick argument may have wider implications in that it suggests that although dynamically the industry will follow a development path, theoretical approaches to dynamic development as put forward by Magee (1976, 1977) and Vernon (1966, 1971, 1974) are simplistic in that they do not take into account the continual internal development and diversification necessary to remain competitive (through comparative advantage) that is likely to be an important factor in most industries, including international construction.

6.1.3 Vertical integration

Although the benefits of vertical integration (VI) in international construction may be eroded by host government protectionist policies (cf Chapter 4), it is still likely that VI is a feature of internal strategy in the industry. The reasoning behind this follows the analysis above concerning HI. VI may be used as an instrument to develop and maintain the competitiveness of a contractor. Given that the competitive nature of the contractor will be a reflection of the country specific comparative advantage of the contractor (as discussed in Chapter 5), then clearly VI strategies will differ according to the nationality of the contractor. We cite here three examples of this:

1. In Korea, VI by the Chaebol has involved backward integration into the materials that are used in the construction process. This reflects the need by the Koreans

to be price competitive because they compete in that part of the market in which price rather than technical expertise is at a premium. This reflects the nature of the Koreans' major country specific advantage (ie cheap labour supplies).

2. Many DC contractors have integrated into construction related services to help them compete on product differentiation against contractors with similar skills. 70% of the survey participants for example said that they have an internal financial department that is involved with foreign currency dealing, bartertrade (where applicable), and project finance which would normally be carried out by separate institutions.

3. Several of the large US contractors have specialist manufacturing concerns used to make machinery necessary for the specialist construction service they provide, including nuclear power plant and petrochemical related equipment.

In each of these cases, VI is the preferred route to the external market because it minimises the potential costs of the market and thereby guarantees the contractor access to resources that may be used to exploit the comparative advantage of the contractor. We would therefore argue that Casson's (1985c) analysis that UK contractors have not followed a comprehensive VI strategy may be correct as far as backward integration into materials is concerned, but fails to notice the service related VI based on comparative advantage which is a central internal motivation for VI. In that this reflects the essence of the Buckley and Enderwick (1985) argument, it is suggested here that this approach may be useful both for strategic planning by contractors in the international environment, and as a valid extension to internalisation theory.

6.2 LOCATIONAL ADVANTAGES

Possibly the most practical way of analysing the locational patterns of international contractors is to separate them into those that affect the location of the individual firm and those that influence the location of a specific nationality group, although in practice the two are likely to be determined by similar locational factors.

Internalisation and Locational Factors

6.2.1 **Locational choice of the individual firm**

Although firm specific locational factors are of interest here, it should be noted that they only form a part of the overall locational criterion of contractors since this will additionally include the interaction of home and host country factors and also take into account O and I characteristics. Hence as we have argued in Chapter 4, while market demand is likely to be a necessary condition it is not generally sufficient to explain locational choice.

Locational choice of the firm is likely to involve three separate decisions:

1. The initial decision to search for overseas opportunities.
2. Choice of market.
3. Extent of involvement in that market.

These are now discussed in turn.

6.2.1.1 **The initial decision**

To illustrate the initial motivation behind searching for overseas opportunities (which is likely to be largely determined by locational factors), contractors in the survey were asked to outline the factors that provided the impetus for entry into overseas markets. Table 6.6, which illustrates the initial incentives to overseas work, suggests that in many cases the decision to go abroad is based upon several considerations. The two major reasons cited in the table indicate that high overseas demand and/or low domestic demand have been the impetus for the majority of contractors to undertake international operations which is clearly in line with the analysis of Chapter 4. Over one-third of the contractors have carried this out through host government invitation or work for home country clients in overseas regions, though only 10% of the total have gone abroad on home government incentives suggesting a low level of government support in this area. As the lowest portion of the table indicates 40% of the survey suggested that further motivation for overseas work has been a means of diversifying risk in addition to higher demand overseas. Overseas involvement to keep up with home country competitors (ie Knickerbocker's oligopolistic reaction) was only considered a relevant motive by 10% of the total while

20% said that they had considered overseas expansion as a conscious policy to maintain dividend remittances to shareholders.

Table 6.6: Initial Incentives to Overseas Work of Participants in Survey

	%
Improve profit in high demand areas	65
Maintain profit at time of low home demand	55
Host Government invitation	45
Undertake work for home country clients abroad	40
To diversify risk	40
To maintain dividend remittances to shareholders	20
Home Government incentive/advice	10
To keep up with home country competitors abroad	10

Source: Ibid.

On this last point, the empirical result goes against that of Neo (1975), who argues that protection of shareholders' interests is a major motivation for international involvement. To verify this, contractors in the survey were questioned upon the importance of shareholders' opinion in the choice of overseas markets. 75% of the total said that their shareholders' opinion was not important so long as dividends did not dramatically fall, while the remaining 25% said that shareholders' opinion was considered when the firm wished to enter a highly unstable or controversial market (eg South Africa). We would therefore argue that the maintenance of dividend remittances to shareholders cannot be classed as a major motivation for overseas expansion according to the survey results.

6.2.1.2 Choice of market

Once the initial decision is taken, the contractor must decide which market to enter. As a means of analysing those factors most important in locational choice, participants in the survey were asked to rate approximately forty potential locational influences according to the significance of each to the contractor's choice of location of office/subsidiary in the Middle East. The ratings were from 1-7; 1-3 were classed as disadvantages with 1 being the most serious disadvantage, 4 corresponded with the factor having no influence on locational choice, and 5-7 represented advantages of locational choice with 7 being the greatest incentive for entering the market. The results were then averaged to give an approximation of the importance of each factor over the survey as a whole. The results are presented in Table 6.7.

Within the table, as the mean and 'influence' columns indicate, there are 20 factors that on average are considered significant advantages (ie over 4.5), 5 significant disadvantages (below 3.5) and 14 factors that are classed as of no significance overall (between 3.51 and 4.49). The greatest motivation for locational choice in the table is size of market, which is in line with the analysis of Chapter 4 and the analysis in the sub section above. In general, most of the other advantages cited require little explanation. Close proximity to offices providing services necessary for the contractor's operations, proximity to other subsidiaries of the firm, availability and quality of expatriate, local technical and local labour manpower, and conditions of the local business framework all confer benefits upon the contractor that would be advantageous in any location where the contractor operates. Possibly the factors of most relevance to locational choice besides market demand (which in the Middle East is likely to have been a major incentive in choosing to enter the market initially) are the financial incentives of the region and home and host country links. These are summarised in Table 6.8, which illustrates the importance of the various incentives according to the mean figure in Table 6.7. The higher the figure in Table 6.8, the greater is the influence of that factor on average to locational choice as suggested by the survey. Financial advantages of locating in the Middle East stem from the fact that in the past the major markets of the Middle East have pursued a policy of low (or zero) personal and company taxation, and non-restrictive repatriation of profits to encourage DC contractors to enter the region. This has clearly been of benefit to the international contractor, so that the freedom on capital movements, when coupled with a high demand situation and low taxation environment are likely to have been the major financial incentives to international contractors, as suggested in Table 6.7. Relationships between the home and host country are also stressed in the table as an incentive to locational choice since these may often determine the level of success or

Table 6.7: Locational Influences on Contractors in Middle Eastern Market

		7	6	5	4	3	2	1	x̄	influence
	Size of regional market	16	3		1				6.7	+
	Number of local contractors	1	1	1	7	4	7		3.3	–
	Number of foreign contractors	1	2	1	1	4	8	2	2.9	–
	Number of home country contractors		2		10	6	3		3.8	0
	Proximity to:									
	Corporate HQ	1		2	16	1			4.2	0
	Offices of Companies providing similar services	1		3	14	1	1		4.1	0
	Offices providing services necessary to company		3	6	11				4.6	+
	Other subsidiaries of the firm	2	4	2	12				4.8	+
	Travelling costs of executives		1	2	16	2			4.3	0
Manpower	Expatriate (work permits)	4	5	3	3	4			4.8	+
Quality/	Local executive/managerial	2	2	5	11	1	1		4.4	0
Avail-	Local professional/technical	2	4	5	7	2		1	4.9	+
ability	Local labour	1	3	3	11	2			4.5	+
	Local imported labour	1	5	3	9	2			4.7	+
	Hiring cost manpower regulations	1	4	1	7	7			4.2	0

Category	Factor								Score	Sign
Manpower Costs	Expatriate costs	1	4	2	3	7	1	2	3.9	0
	Local executive/managerial	2	2	1	15	2			4.2	0
	Local professional/technical	2	2	2	8	4	2		4.2	0
	Local skilled/unskilled	4	2	1	8	4			4.7	+
	Manpower costs regulations	2	3	1	8	5		1	4.2	0
Social Conditions	Language	2	4	2	6	3	2	1	4.3	0
	Expatriate social and living conditions	2	3		8	3	2		4.3	0
	Distance from home country	2	2		15	3			4.0	0
Business Framework	Availability of financial and consulting services	2	2	3	13				4.6	+
	Personal tax level	2	7	2	9				5.1	+
	Corporate tax level	2	5	4	7	2			4.9	+
	Availability of host government incentives	1	3	5	10	1			4.6	+
	Attitude of host government to company	3	8	5	3				5.3	+
	Attitude of clients to company	3	9	6	2				5.6	+
Trading Freedom	Controls on capital, imports, dividend remittances etc	7	10	3					6.2	+
	Tariff and import controls, immigration regulations	3	8	4	4	1			5.4	+
	Non-tariff barriers eg. host government procurement policies	2	2	2	6	6	1		4.0	0

Table 6.7: continued

	Home government incentives		2	1	15	1		4.1	0
	Home/Host country political links	5	5	6	4		1	5.5	+
Home Country	Long association of firm to region	3	6	5	6			5.3	+
	Mutual government agreements (eg single taxation)	1	6	9	4			5.2	+
	Number of home country companies (not construction) in region	2	9	8	1			4.5	+
Political Environment	Possibility of revolution		1	4	4	7	5 3	2.8	−
	Possibility of expropriation of firm's assets		1	1	4	8	4 2	3.0	−

Source: Ibid.

Internalisation and Locational Factors

restrictions the contractor will face. It should be noted that in Table 6.7 home government incentives are not mentioned as a significant influence in this respect since in most cases these did not exist for survey participants. Nevertheless, locational choice does appear to be significantly influenced by political links and the local attitude towards the contractor which are likely to reflect country specific factors.

Table 6.8: Major Incentives of Location by Mean

	$\bar{x}$
Size of regional market	6.7
Controls on capital, imports dividend remittances etc	6.2
Attitude of clients to company	5.6
Home/host country political links	5.5
Tariff and import controls, immigration regulations	5.4
Attitude of host government to company	5.3
Long association of firm to region	5.3
Mutual government agreements (eg single taxation)	5.2
Personal tax level	5.1
Corporate tax level	4.9
Local professional/technical manpower quality	4.9
Proximity to other subsidiaries of the firm	4.8
Expatriate availability (ie work permits)	4.8
Local skilled/unskilled manpower costs	4.7
Local imported labour availability/quality	4.7
Availability of financial and consulting services	4.6
Proximity to offices supplying services necessary for close operations	4.6
Availability of host government incentive	4.6
Number of home country firms (not construction) in region	4.5
Local labour availability/quality	4.5

Source: Table 6.7

Disadvantages associated with locational choice in the table fall into two distinct groups, those relating to competitors' operations and political risk. Table 6.9 gives a summary of these factors on a similar basis to Table 6.8, except that in

Table 6.9 the lower the figure the more of a disadvantage the factor is to locating in that region. In terms of competitors' operations, the contractor is likely to favour a market with relatively few competitors (foreign or local) because of the determinants of demand and supply and the structure of the industry as argued in Chapter 4. Additionally, political risk may be considered a disadvantage to the contractor because it may put the firm's assets at risk or increase the possibility of default on payments to the contractor.

Table 6.9: Major Disincentives of Location by Mean

	$\bar{x}$
Possibility of revolution	2.8
Number of foreign contractors	2.9
Possibility of expropriation of firm's assets	3.0
Number of local contractors	3.3

Source: Table 6.7.

Of the factors not considered relevant, manpower costs in general were not regarded as significant determinants of locational choice mainly because all firms have to pay local and expatriate costs (if they have no source of cheap labour supply), so that this is unlikely to affect the contractor's cost situation in relation to competitors in the same situation. It is interesting to note that in terms of manpower quality the only factor not considered important in the table is quality of local management, indicating that managerial and executive functions are carried out by the firm's personnel so that local management and executive manpower are not required. Travelling costs of executives are not significant because they form a minor part of costs that will be covered by the price, while proximity to the corporate HQ is not considered important because of the mobile nature of the contractor's product that can act autonomously when required. None of the social conditions in the table were considered important to locational choice, and while language was expected to have some influence on locational choice the widespread use of English in the Middle East may have reduced the significance of this factor to the survey participants, the majority of which speak English as a first language. The common use of English in the Middle East is likely to confer an O advantage

Internalisation and Locational Factors

upon English speaking contractors in the region.

6.2.1.3 Determination of extent of involvement

In addition to deciding which market to enter, the contractor must decide what level of involvement to have in a market which ultimately comes down to the decision between setting up a temporary or permanent presence in that country. In practice this decision is likely to take place over time as more about the market is learnt and predictions of the future become more accurate (Johanson and Vahlne, 1977). Table 6.10 summarises those factors considered relevant in this choice given by the survey respondents. Clearly the most important factor in the decision is the level of potential and/or actual market demand, which gives empirical justification to market demand being the determining force in the Buckley and Casson (1981) model as outlined in Chapter 4. However, from Table 6.10 it is obvious that other factors are also taken into account. The table suggests that these reflect the major locational determinants discussed in Tables 6.7-6.9 that relate to political stability, financial returns and host country relations with the firm. It is interesting to note that level of competition is not regarded as a major criterion in the table although in practice this may be classed as a component of demand so that the contractor is not likely to enter saturated or well established markets where there is a high level of competition because effective demand for the contractor's product is likely to be low.

Table 6.10: Factors Considered Important as Criteria of Temporary/Permanent Subsidiary set up from Survey Results

	%
Market potential and/or actual demand	85
Remittance of profit possibilities	35
Political stability	25
Host government policy towards the firm	25
Local attitude to the company	20
Level of competition within the country	10
Social atmosphere	5
Labour pool and quality	5

Source: Fieldwork

Internalisation and Locational Factors

Figure 6.5: Location of Offices/Subsidiaries and Construction Works Carried out of Survey Participants*, 1982

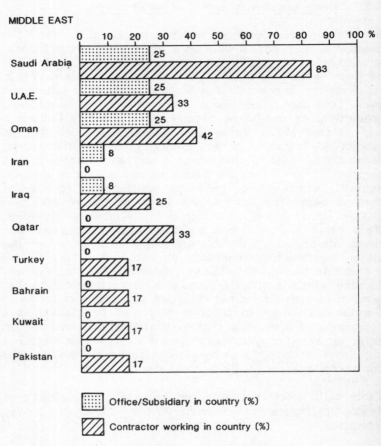

NB: Not all participants included due to insufficient information

Internalisation and Locational Factors

LATIN AMERICA

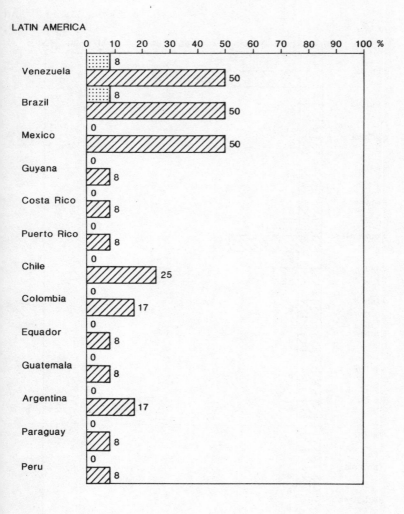

Internalisation and Locational Factors

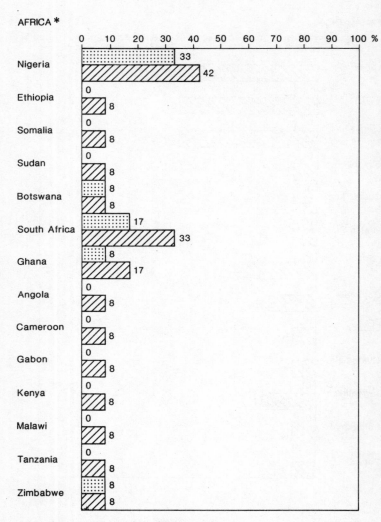

*(excluding North Africa, Egypt, Algeria, Libya, Tunisia)

Internalisation and Locational Factors

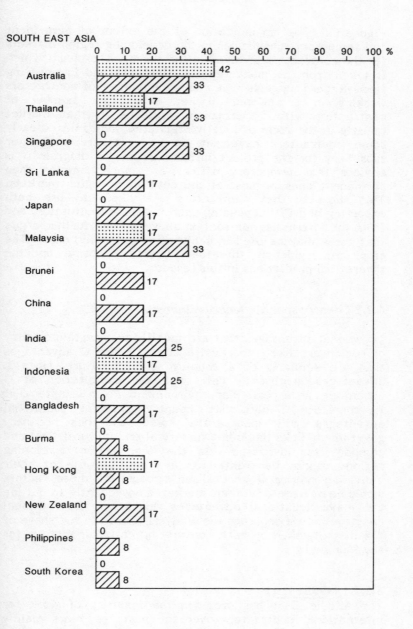

Figure 6.5 gives an indication of the extent of permanent and temporary subsidiaries in the developing region markets for the contractors involved in the survey. The chart, which is taken from company reports, fieldwork and Engineering News-Record suggests that where the number of contractors working in a national market exceed the number of contractors with a permanent presence in that market (mostly in the form of locally registered companies) clearly some contractors have set up a temporary office or subsidiary for the project in hand. From the diagram it is obvious that temporary offices are more common than permanent ones in most of the developing region markets. This suggests that contractors are likely to use both exporting and FDI depending upon the market situation, and from the internalisation section above we can further argue that these may be used in conjunction in certain cases. The empirical evidence therefore tends to back up the theoretical predictions in this respect.

6.2.2 Country specific location factors

Locational advantages that are specific to a certain country or region are likely to revolve around those O advantages that are similarly of a country specific nature, since L advantages specifically rely upon the exploitation of O advantages in a particular environment. From Chapter 5 it is therefore likely that country specific locational advantages will incorporate home and host country government links (including historical and cultural ties) and comparative advantage. As the four major developing regions cannot be treated as a group because of their individual nature a generalised account of these factors would be of little use in any market. Consequently to outline the major locational advantages relevant in international construction we separate the analysis to isolate the share of the market taken by each contracting nationality group for the four regions.

6.2.2.1 Middle East
The Middle East has provided the majority of work for international contractors over the past ten years mainly because of the high level of effective demand that resulted from the increased revenue from oil following the

quadrupling of oil prices in 1973. However, in recent years the decline in oil revenues and the completion of major infrastructure projects have had a limiting effect upon the rate of growth of construction demand in the Middle East.

The major features of the Middle East that provide the basis for exploitation of comparative advantage are that the region depends heavily upon the production of oil for revenue, lacks both skilled and unskilled labour and generally is sparsely populated. The consequences of this are twofold:

1. Because of the limited supply of labour, construction projects in the Middle East are primarily energy and capital intensive so that DC contractors have benefited overall from the construction boom. Additionally, the dependence on oil has led to the need for extensive construction in oil related fields such as petrochemical plants so that high tech DC contractors have benefited here also.

2. Connected with 1, the lack of labour in the region has meant that the high demand for workers needed to carry out construction has led to an enormous migration of workers from the LDCs and NICs, where there are plentiful supplies of cheap labour, to the region.

Given these conditions, some nationality contractors have found the market easier to compete in than others because of the comparative advantage they possess. Table 6.11 illustrates the major contractors' awards in the Middle East over the period 1980-84. The 'total' figure shows that the market peaked in 1982 and has fallen since, and that this pattern has been reflected in the earnings of American, German and South Korean contractors in the Middle East although in 1984 the Japanese and French also suffered a fall in awards. The UK contractors have experienced a rising share of the market over the period, though they have not seen as dramatic a rise in earnings as Turkish contractors who have increased awards in the market tenfold over 1980-84. Italian contractors have taken an insignificant share of the market over the period analysed possibly because of the concentration of Italian interests in other regional markets.

Because of the need for high tech construction in the Middle East, particularly in oil related fields, American contractors have benefited the most as a result of their comparative advantage in this respect (cf Chapter 5), and to a lesser extent the French. The success of the Koreans in the region is mainly attributable to the labour force the

Table 6.11: Share of Middle Eastern Market by Country of Origin of Contractor, 1980-85

Country	1980	1981	Value of Work Done (£m) 1982	1983	1984
U.S.	4,531.38	7,452.50	10,504.74	8,407.65	8,004.19
Japan	349.28	508.61	255.28	438.63	318.63
France	N/A	N/A	2,617.80+	3,559.61	1,439.22+
U.K.	612.00	552.00	651.00	736.00	741.00+
Germany	1,582.30	2,037.60	1,432.92	527.59	196.22
South Korea	3,337.69	6,273.83	6,500.31	5,999.64	4,419.70
Italy*	N/A	157.08	4.25	1.38	1.52
Turkey	444.54	1,418.50	2,360.13	2,157.22	4,268.76
Total:	10,847.19	18,400.22	2,364.38	2,158.60	4,270.23

+ Estimated
* Middle East here classed as Abu Dhabi, Saudi Arabia, Iran, Iraq, Kuwait.

Sources:
US Engineering News-Record (various issues)
Japan Japanese Ministry of construction 'Japanese Overseas Construction' (July 1985)

Table 6.11: continued

France	Federation nationale du Batiment 'Travaux de bâtiment à l'etranger en 1983' (January 1983) European community contractors 'Construction in Europe Statistics' (1984)
UK	Department of Environment 'Housing and Construction Statistics' (1985)
Germany	Hauptverband der Deutschen Bauindustrie 'Baukonjuntur – Spiegel, (January 1985)
South Korea	Construction Association of Korea 'The construction Association of Korea' (1985)
Turkey	TUSIAD 'The Turkish Economy 1984' (1984)
Italy	ANCE 'L'industria delle costruzioni nel 1984' (May 1985) CIM 'Costruttori Italiani nel Mondo' (March 1985)

Korean contractors export for construction in the Middle East. The Koreans are not only noted for their diligence and hard work, but also the fact that they keep themselves separate from the local population and do not cause immigration problems since they leave the country once the project is completed. This factor is becoming increasingly relevant in the Middle East where there is growing concern because of the rising number of illegal immigrants that settle permanently in the region in search of work. In many cases this has resulted in public authorities hiring Korean contractors on public construction works to minimise immigration problems: '... the fact that Korean labourers tend to keep themselves apart and return to their native land after the completion of the contract is an additional advantage.' (Financial Times, 22/1/80). Hence the Koreans have concentrated in the Middle East because this enables them to exploit their major O advantage to the full given the nature of the location.

Other advantages to contractors in the region may exist because of cultural or historic links. The Turkish contractors' major advantage in the region for example is that Turkey is a Moslem country and considered local by many of the Middle Eastern states so that protectionist policies and local bias tend to favour the Turks against other nationality contractors. This may explain the growth of the Turk's share of the Middle East market over the period, and is likely to be the motivation behind the Turkish contractors concentrating in the predominantly Moslem Middle East and North African regions and largely ignoring the South East Asian and Latin American markets. Host and home government links also appear to be a factor that have determined locational decisions in the past. Many UK contractors have been successful in Oman for example because of close links between the UK government and Oman on aspects of trade and defence. A similar arrangement has benefited US contractors in Saudi Arabia, although this link has been eroded somewhat in recent years because of American involvement in Israel. Further evidence of the impact of government influence affecting contractors' operations is that of the South Korean government given in Chapter 5, who have negotiated several government to government deals in the Middle East to provide construction services in return for oil. This 'double coincidence of wants' (ie the Koreans require oil and have the construction services while the Arabs have the oil and

Internalisation and Locational Factors

need the construction) has been an instrumental feature in the maintenance of Korean contractors' workload in the region despite dwindling demand for foreign contractors' services.

The greatest locational advantage within the market then is that the Middle East has enjoyed large sustained demand for construction services, but at the same time has not had a competent indigenous construction sector to supply the required services. These factors together have provided the major impetus to the influx of contractors from LDC, NIC and DC regions. This has led to a high level of competition in the region whereby the contractor has had to rely upon product differentiation according to the nature of the O advantages the firm possesses. This situation is likely to become even more prevalent as demand for construction works falls and indigenous contractors reach a level of expertise that is competitive with the foreign contractors presently in the region.

6.2.2.2 South East Asia

Many contractors in the survey when questioned about future prospects for the industry considered Asia, particularly South East Asia, as the market with the most potential of all developing regions. This is largely based upon the high rates of growth many of the South East Asian countries have achieved over the 1980s. From 1980-83 South Asia had an annual average real rate of economic growth of 5.5% while East Asia and the Pacific had a 5.6% growth rate. This is comparable with an average rate of 2.3% for all developing countries, and 1.1% growth in industrial market economies ('World Development Report 1984', World Bank 1984).

Possibly the most important factor in the expectations of construction companies in the Asian market is the potential that China has to offer in the form of extensive capital expenditure programmes aimed at pushing China into the industrialised world. The nature of this programme and the size of the country ensure that this is likely to require extensive construction for both social and industrial development, costing billions of Dollars. However, other countries such as Thailand, Malaysia, Indonesia, the Philippines, Singapore and Brunei also offer promising construction markets based upon the need for infrastructure development and improvement and the willingness of the

Table 6.12: Share of South East Asian Market by Country of Origin of Contractor, 1980-84

Country	1980	1981	Value of Work Done (£m) 1982	1983	1984
US	3,654.34	5,145.86	5,098.63	3,262.42	6,611.08
Japan	462.28	1,154.15	1,659.35	1,900.75	1,720.58
France	N/A	N/A	143.41	194.10	812.37
UK	89.50	96.50	180.50	285.00	306.00+
Germany	0*	0*	0*	0*	0*
South Korea	142.03	269.84	994.16	688.46	428.34
Italy	N/A	466.74	385.31	63.98	105.07
Turkey	0*	0*	0*	0*	0*
Total	4,348.10	7,133.10	8,461.40	6,394.70	10,164.20

* Less than £0.5m
+ Estimated

Source: Ibid.

Internalisation and Locational Factors

multilateral lending agencies to provide the finance to achieve these goals.

Table 6.12 illustrates the share of the Asian market taken by the major contracting countries for the period 1980-84. The market has experienced an overall rise in the level of awards over the period despite a minor fall in 1983 which may reflect adverse political conditions in Indonesia at the time. The dominance of construction activity in the region by American and increasingly Japanese contractors reflects the demand for large scale industrial construction (where both sets of contractors have an advantage), political relationships and especially in the Japanese case, the provision of bilateral aid and grants for development that is often 'tied' so that all services and materials used by the borrower country for construction come from the loaner country. This subject is discussed in more detail in Chapter 7.

Both the French and UK contractors have increased their market share in Asia over the period, though neither has a large proportion of the market when compared to the Japanese and American contractors. German contractors had not had a significant involvement in the market until 1984, and this move to the Asian market may have possibly been to offset German losses in the Middle East as illustrated in Table 6.11. Italian contractors have faced a spasmodic market for their services in South East Asia, as shown by the fluctuating level of awards in the region. Surprisingly, like the Europeans, South Korean contractors have only taken a minor portion of the market which is thought to reflect the concentrated effort of Korean contractors in the Middle East. Over the period 1980-84 Middle Eastern awards of Korean contractors accounted for approximately 90% of average yearly overseas earnings, while Asian awards contributed 8% of the total. This clearly illustrates a conscious policy in terms of locational concentration on the part of the Koreans. The fact that Turkish contractors were not represented significantly enough to be represented in Table 6.12 suggests that the Turks have followed a policy similar to that of Korean contractors which is likely to have been based upon exploitation of O advantages.

Possibly the major contracting countries with the greatest implicit advantage for exploiting the developing South East Asian market are the Japanese and Koreans because of cultural links and proximity to the market. The potential for all international contractors may be constrained however by the fact that in several of the countries indigenous construction sectors have expanded and developed to the extent that some contractors from the region are represented in the annual ENR survey of the top 250 contractors. Generally however, these rely upon construction skills similar to those of the Koreans and depend upon a plentiful supply of cheap labour from the home country to compete abroad. This may provide an explanation for the low level of Korean involvement in the region.

While the Americans have strong political links with several South East Asian countries, and the Japanese and Koreans can exploit advantages of proximity and cultural similarities in the region, the European contractors as a whole are not well represented in the South East Asian market. It is possible that UK contractors may have a foothold in the market because of political and historical connections with Hong Kong, but with a highly competent indigenous construction sector in the province this should not be overstressed. If we additionally consider that many of the indigenous construction sectors in the region are rapidly accumulating experience and increasing the level of their construction skills, then it is clear that in the future competition in the region is likely to be severe for all but the international construction firm with a specialist market that can compete on a high level of technical ability.

6.2.2.3 Africa

Although over the last five years market awards of the major contracting countries have risen in the African market (with the exception of 1984, as Table 6.13 illustrates), the majority of the demand for construction in the region has come from the northern African states of Libya, Egypt, and Algeria while sub Sahara Black Africa especially has had to struggle to maintain past levels of work. This is mainly a consequence of the global recession that has led to high foreign indebtedness so that in many countries even concessionary finance from multinational aid agencies is declining. Additionally, many of the countries are considered politically unstable so that bilateral finance is only available for those markets that show relatively high revenue, which in the main are the oil producing and exporting countries of Nigeria, the Ivory Coast, Cameroon,

Internalisation and Locational Factors

and the Congo. Nigeria is the largest construction market in Black Africa.

Table 6.13 shows the awards of the major contracting groups in the region over the period 1980-84, and illustrates that Turkish, American, and French contractors have been especially successful in Africa, although it should be stressed that the Turkish awards were won in North Africa only (especially Libya) where the Turks have been successful because of political and cultural links similar to those given above with reference to the Middle East. The Italians similarly have concentrated their African operations in North Africa, (also mainly in Libya where the demand for construction is high) and have been successful generally through government to government negotiations and because of the small number of other European and American contractors in Libya due to the political sensitivity of the country.

As in the Middle Eastern and Asian markets, American contractors seem to have been heavily in demand because of the high tech nature of the US contractors' products, while the success of French contractors in the region may be considered both a consequence of historical links and strategic policy. French contractors have been advantageous in that Francophone Africa, (which is a product of the French empire in Africa), as the name suggests has evolved the same legal and technical standards frameworks as France, and additionally uses French as the native language so that other nationality contractors face a language and technical barrier to entry of these markets that obviously benefits French contractors. In addition to this however, the traditional British markets in Nigeria and East Africa have been challenged by French contractors from their bases in French West Africa by hiring British consultants and specialists to deal with any barriers the French may face in the British ex-colonial states. The tactic appears to have been successful - in 1982 33% of French contracts in Africa were from English speaking countries (Financial Times, 5/3/84). As a result of this aggressive policy UK contractors have faced an uncertain market in Africa while the Germans, who have also regarded Africa as a traditional ex-colonial market in the past, have experienced a falling share of the region's awards. Competition in the African market has not just come from the French however. Japanese contractors have entered the market with home government help in search of natural resources, while South Korean

219

Table 6.13: Share of African Market by Country of Origin of Contractor 1980-84

Country	1980	1981	Value of Work Done (£m) 1982	1983	1984
US	1160.79	1869.43	1640.91	1623.73	1258.83
Japan	71.91	78.25	21.27	87.73	31.86
France	N/A	N/A	1816.90	1321.00	1533.42
UK	217.00	252.00	828.00	546.50	569.00+
Germany	908.08	656.42	410.81	228.90	120.50
South Korea	0*	0*	0*	0*	0*
Italy	N/A	840.16	559.62	532.88	673.05
Turkey	1067.50	3045.00	4793.59	8235.75	6488.03
Total	3425.30	6741.30	10071.10	12576.50	10674.70

* Less than £0.5m
+ Estimated

Source: Ibid.

contractors are increasingly turning to the market as Middle East contracts fall off, though with a large labour sector in most of Africa the Korean success in this region may be limited. Nevertheless, both could present a major problem to established contractors in the market region, that is characterised by a highly competitive environment despite the relatively low level of demand overall.

6.2.2.4 Latin America

Although the Latin American market is one with large potential for construction works, it is also a region with serious financial problems that have had a marked effect on the effective demand for construction. A high level of indebtedness by most countries in the region to multilateral agencies and DCs, coupled with inflation that is generally over 100%, has led to a combination of austerity measures, cuts in development programmes, reduced foreign investment, and strict import controls which have economically depressed the area, and consequently the construction environment. Of the four developing markets illustrated, Latin America is the smallest. This is illustrated in Table 6.14, which shows awards of the major contractors in Latin America over the period 1980-84, although the totals for 1980 and 1981 are underestimated by the lack of data on the Italian, French, and British contractors' earnings in the region for these years.

Table 6.14 shows that the market peaked in 1981 and has declined since, partly through self imposed economic restraint but also because of the effects of global recession in the region. American contractors are the clear market leaders, which is possibly as attributable to the proximity of the market to America and the level of US financial help to several Latin American countries as the skills of the American contractors in capital intensive construction.

While the Japanese, British, and German contractors have not fared especially well in the region, Italian and French contractors have been relatively successful (though the French have faced an unsteady market as shown by the fluctuation of the awards over the period), mainly because of aggressive market policies by both, and possibly cultural and language links on the part of the Italians. The absence of South Korean and Turkish contractors confirms that both have concentrated in the Middle East where their O advantages can be fully exploited, which would not

Table 6.14: Share of Latin American Market by Country of Origin of Contractor 1980-84.

Country	Value of Work Done (£m)				
	1980	1981	1982	1983	1984
US	3215.82	5399.31	2566.34	1146.47	1226.72
Japan	92.46	156.49	63.82	58.48	95.59
France	N/A	N/A	249.52+	50.14	270.63+
UK	N/A	N/A	70.80+	183.40+	170.07+
Germany	88.36	88.09	96.31	124.68	40.71
South Korea	0*	0*	0*	0*	0*
Italy	N/A	634.20	45.47	370.07	154.32
Turkey	0*	0*	0*	0*	0*
Total	3396.62	6278.09	3092.26	1933.24	1958.04

* Under £0.5m
+ Estimated

Source: Ibid.

necessarily occur in Latin America. Unlike the sparsely populated Middle East or managerially unskilled African market, Latin America has a construction sector that is both skilled and competent so that foreign contractors are expected to make maximum use of local resources, including the large indigenous labour sector. This obviously would inhibit the South Koreans from exploiting their major O advantage, and may be instrumental in explaining the lack of Korean earnings in the market. For DC contractors, the fact that local contractors are already skilled implies that joint venture may not expose the foreign contractor to a high level of risk to firm specific assets, but helps minimise risk exposure in what is a politically risky area. DC contractors may therefore find that the level of competence of local contractors in the Latin American market is beneficial to their operations in the market.

6.2.3 The role of the Kojima argument in international contracting

Kojima's (1978) argument, that suggests that Japanese enterprises use FDI as a tool for technology transfer to exploit comparative advantage of both the domestic and foreign markets, may have some relevance to the overseas operations of the Japanese contractors and also the Koreans who follow a similar strategy in this respect. From interviews with people connected with the industry it seems that Japanese and Korean contractors use joint venturing as a means of technology transfer to lesser developed countries, while retaining a comparative advantage in skills that cannot be embraced by those contractors. In both cases, government policy appears to be the motivation behind this type of involvement. The Korean government for example recommends Korean contractors to joint venture wherever possible with local contractors to gain local knowledge of the market and encourage technology transfer to lesser developed contractors (Sung Hwan-Jo, 1982), and additionally encourages contractors to reinvest profits in these countries rather than repatriate the returns for several reasons:
  1. Profits can be invested in industries providing raw materials such as coal and steel that can then be exported to South Korea for use by domestic enterprises.
  2. Conglomerate investment abroad enables Korean

contractors to have a diversified equity base that will provide continuous income for the firm when construction demand overseas is in recession.

3. The conversion of foreign currencies in large amounts creates inflationary pressures in the domestic economy and therefore should be avoided.

Of particular relevance here is the motivation to acquire raw materials for domestic use, which is also important to Japanese contractors' overseas operations. In return for joint venturing the Koreans and Japanese are frequently paid in raw materials, which they use for domestic industry and which are necessary since both Japan and Korea have very limited supplies of natural resources that can be used as raw materials in manufacturing. In both cases therefore, the interaction of locational factors of the home and host country provide justification for entering certain markets, such as the Middle East with extensive oil reserves of certain African markets that are well endowed in natural resources, and also for an integrated VI policy as already discussed, to develop comparative advantage. Because technology transfer provides the basis for the acquisition of these materials, it appears that Kojima's Japanese style FDI may be empirically valid in international construction. However, there remain serious flaws in the approach:

1. Although Kojima's 'Japanese type' FDI is represented in the industry, there is no reason to argue that the Kojima 'American style' trade inhibiting type FDI is also present. From the discussion of the various market regions above it is obvious that US and other DC contractors inhabit the same markets as the Japanese, and are often involved in joint ventures which are likely to include some aspect of technology transfer. Comparison of the trade inhibiting and trade creating types of FDI are therefore not valid within the industry so that the Japanese style FDI is not necessarily more efficient.

2. The fact that Japanese contractors receive 'payment' for technology transfer in raw materials for use solely in Japanese industries may be criticised on political grounds because it does not seem far removed from European colonial links of the 19th century, and on economic grounds because the policy replaces the external market with a type of internal link between the producer and the user which may reduce the benefits the producer of the materials may enjoy in the external market. It is unclear

Internalisation and Locational Factors

how this may be regarded as economically efficient by Kojima, when in reality it may be the opposite for the producer of the raw material.

On the basis of these criticisms we would argue that while Kojima may be justified in suggesting that Japanese industry may follow a pattern similar to that outlined in the theory, the motives and consequences of this strategy may not be so readily accepted.

6.3 POLITICAL RISK IN INTERNATIONAL CONTRACTING

Although internalisation may reduce risks of an inefficient or missing external market with regard to production supplies or the ability to maintain a competitive strategy against aggressive competitors (Graham 1974, 1978, 1985), Casson (1979) argues that this increases the possibility of loss of assets or O advantages through xenophobic or nationalistic tendencies by both clients and governments in foreign markets. This is considered a major potential cost of the internal market, as discussed in Chapter 2. To prevent this internal cost from occurring, Vernon (1983) suggests that partial externalisation may lessen the exposure to risk that the MNE faces in an unstable market. Vernon puts forward three possible options to lessen risk exposure:

1. Foreign consortium. By joining a foreign consortium the contractor can work in a country with only a low level of direct exposure to the firm, allowing the company to utilise those assets saved by entering the consortium in other markets. However, this only reduces the potential loss rather than the level of risk, and may not be acceptable to the MNE for reasons that relate to the returns the contractor gets (which are inevitably smaller than if the contractor carried out the project alone), and internal efficiency of the consortium.

2. Joint venture. By joint venturing with either a local company or a state owned enterprise the MNE will face less nationalistic pressure and thereby lessen political risk. In many LDC markets joint venture with a local company is required by law, although as with a consortium internal efficiency of the venture and ability to maintain O advantages may present a problem to joint venture.

3. Other measures. The two other major ways the MNE can lessen risk exposure are bribery of officials and nationals in the host country to gain favour, and only

undertaking long term contracts in the market so that fixed assets are minimised. Although neither can prevent default, a long term contract is probably the most predictable of these, and certainly the easier to insure against default.

Within the Vernon analysis it should be noted that the three options are not mutually exclusive, although generally consortia and joint venture will not exist for the same project and same partners.

When applied to international construction, it appears from the results of the survey that Vernon's analysis may be justified. Of the 100% of the survey that said they would joint venture, 60% of the total regarded diversified risk in politically unstable areas as a major advantage of joint venturing. Although Vernon does not outline the internal problems that the MNE may face when joining a joint venture or consortium (in particular the maintenance of the company's O advantages as discussed in the internalisation section above), from the discussion of joint venturing in this chapter and Chapter 5 we would argue that Vernon's analysis successfully summarises some of the risk minimisation policies that are open to international contractors, and gives an insight into the high incidence of joint venturing within the industry, particularly in the politically unstable areas of the world.

A second option open to the multinational firm, which may additionally incorporate the Vernon analysis, is that to minimise risk the firm should follow what MNE theorists refer to as a 'global strategy'. Kogut (1985) suggests that the essence of the MNE's global strategy must be flexibility of operations of the company which enables the MNE to exploit the uncertainty that arises in markets that may be caused by exchange rate movements, competitive moves, or government policy (ie political risk). The thesis of the global strategy is that the MNE must centralise its operations so that subsidiaries act as part of the overall corporate hierarchy to capture transactional advantages available to the enterprise. Whereas Vernon advocates a partial externalisation, the global strategy requires complete internalisation of overseas markets to centralise the overall objectives of the company. In Kogut's analysis this requires the centralisation of six tasks for the firm to benefit from global strategy, four of which are arbitrage opportunities and two of which are leverage points to enhance the strategic position of the firm vis-à-vis competitors and host governments.

1. Arbitrage opportunities. These may arise from the ability of the MNE to exploit movements between national markets and may be classed as transactional advantages in the eclectic framework. In the Kogut model these are:
   a. the shifting of production to take account of national differences in wage rates, resource costs and market conditions;
   b. tax minimisation which exploits transfer pricing and the establishment of multiple channels for income remittances (where tax levels may differ according to the nature of the remittance);
   c. financial market arbitrage to enable the MNE to have access to finance from international sources over and above the national firm;
   d. information arbitrage to match information from different markets to get a clearer picture of overall market conditions.
2. Leverage opportunities are:
   a. global coordination of production to coordinate all pricing and competitive strategies in national markets where competitors produce;
   b. political risk minimisation by being able to increase bargaining power in any national market where competitors produce.

The implication of this strategy for the international contractor is that because the contractor faces a situation that is characterised by a high level of risk (both political and financial), a centralised global strategy may be a means to reduce the overall exposure of the contractor to exogenous variables. For the international contractor, possibly the major feature of a centralised global strategy may be the ability of the firm to coordinate resources to allow the continual use of human resources in major markets via pure exporting or simultaneous exporting/FDI as discussed in the internalisation section of this chapter. By undertaking such a strategy the contractor is not only in the position where Kogut's six advantages may be exploited, but additionally lessens risk exposure to political instability or volatile demand conditions in any national market. Where the firm undertakes pure exporting or simultaneous exporting/FDI the contractor needs only a low level of fixed assets in any country which can be 'topped up' by highly mobile human capital when required. Risk exposure is therefore lessened in any unstable market by coordinating resources across markets, which is the central feature of the global strategy.

From the results and analysis in section 6.1.1 on the internal hierarchy in international contracting firms it appears that empirically there is justification for suggesting that contractors do carry out global coordination which may improve the overall efficiency of the firm in areas where market demand and the political environment are unpredictable. The ability of the contractor to carry out this strategy ultimately comes down to the similarity of exporting and FDI in international construction that is a consequence of the product the contractor offers. As a means of testing further how contractors' operations differ between politically stable and unstable markets, participants in the survey were asked to outline their company policy with respect to 'politically unsound' countries. The results are summarised in Table 6.15. The majority of the contractors in the survey (65%) said that they opt for minimising exposure in politically unstable countries either by keeping only a low level of fixed assets in that country, or maintaining all local financial involvement at a low level. 20% of the contractors embargo politically unstable countries (and one actually undertakes detailed political risk analysis on a formal model to judge which countries should be avoided), while 10% demand prepayment and immediate remittance of earnings, which may be classed as a form of minimising financial involvement. 5% of the total enter unstable markets with extensive risk assurance, although most contractors will take out appropriate risk assurance that is generally provided by the home government (cf Chapter 5).

Table 6.15: Political Risk Measures taken by Contractors in Survey

	%
Minimise fixed assets	35
Minimise financial involvement	30
Embargo certain countries	20
Demand pre-payment and repatriation agreements	10
Obtain adequate insurance	5

Source: Fieldwork.

In summary, the empirical results tend to substantiate the validity of the theoretical works of both Vernon (1983) and

Internalisation and Locational Factors

Kogut (1985) to international construction firms. In practice many contractors do appear to undertake joint venturing as a means of diversifying risk, while additionally the ability of the contractor to exploit uncertainty via the global coordination of resources (as suggested by Kogut) appears to be empirically justified by firms in the survey. (2) This is illustrated by the predominance of exporting and exporting/FDI in politically unstable areas, and the policies of contractors to minimise fixed resources and assets in those markets. Ultimately this may lead to coordination of the internal hierarchy which bestows additional transactional advantages on the enterprise and suggests that the globally coordinated firm will not only benefit in unpredictable markets, but in all the firm's operations worldwide as a result of the adoption of a global strategy. This is clearly a benefit of the internal hierarchy not available in the market and therefore fits into the analysis of Buckley and Casson (1976), Casson (1979, 1982), and Dunning (1981, 1985) as motivation for the internal organisation of the MNE.

6.4 OLIGOPOLISTIC REACTION IN INTERNATIONAL CONSTRUCTION

The empirical analysis presented in this chapter tends to substantiate the theoretical reasoning in Chapter 4 that oligopolistic reaction is not likely to provide motivation for entry to foreign markets in international construction as suggested by the analysis of Knickerbocker (1973) or Graham (1974, 1978). Because oligopolistic reaction is likely to affect O, L, and I factors, empirical analysis of the importance of oligopolistic strategy in international construction was not treated directly in the questionnaire but incorporated into several questions to determine whether this type of strategy had different effects on the components of the eclectic approach. The argument that oligopolistic reaction is not of great importance in international construction therefore comes from several sources:

1. When questioned about motives for establishing a local presence in the Middle East only 20% of the survey mentioned that competitors setting up local operations was significant in this decision, as given in Table 6.3.

2. In Table 6.7, which illustrates factors significant to

locational choice, the number of foreign contractors in a market was considered a disincentive to enter that market, while the number of home country contractors in the market was not considered important to the decision. If oligopolistic reaction were relevant, both would be considered an incentive to enter that market according to Knickerbocker's 'bandwagon' effect.

3. Only 10% of the survey said that an initial incentive to overseas work was to keep up with home country competitors. This clearly invalidates Knickerbocker's approach in this case, which would suggest that a greater percentage would regard this as a major incentive to overseas work.

When taken together, these empirical results provide justification for rejecting the importance of oligopolistic strategy as a major factor in the international operations of contracting firms which follows the theoretical predictions of Chapters 3 and 4. The rejection of Knickerbocker's thesis of oligopolistic reaction in the international construction industry follows the theoretical reasoning of Casson and Norman (1983). Because the international contractor operates in a large international market with a product produced from reasonably standardised human technology (even though there remain differences between DC and LDC human capital) barriers to entry are not likely to be effective so that oligopolistic strategy will have little effect in the industry. Hence the validity of the Knickerbocker approach can be refuted both theoretically and empirically within the international construction industry. Given that the empirical results validate the theoretical predictions concerning oligopolistic strategy when applied to the specific nature of international construction, this further provides support for the use of the theoretical works of Norman and Casson (1983) and Casson (1983) for predictive reasoning of the level of oligopolistic reaction in any multinational industry.

6.5 FOOTNOTES

1. This may not be considered as the usual form of pure exporting in international production terms (ie where there is no value added in the market) but is necessarily redefined here because the mobility of the production process rather than the product in international construction

Internalisation and Locational Factors

implies that the usual definition of exporting is redundant in this case.

2. Although global strategy involves centralisation of the firm's operations, being a part of a joint venture or consortium may be acceptable to the firm if the arrangement allows enough flexibility for the enterprise to additionally carry out global coordination. Where this is not possible however the MNE is unlikely to be involved in both types of risk minimisation.

Chapter Seven

THE FINANCING OF INTERNATIONAL PROJECTS (1)

The factor which has possibly had the greatest impact on competition within the international construction industry in the 1980s is the increased need for finance in developing countries to carry out construction projects. In the past, major sources of these funds have been direct loans by the commercial banks and international development agencies to these regions, but the reluctance of commercial banks to lend to the highly indebted and often politically unstable developing areas and the limited capacity of the development agencies to provide development finance has placed the burden of provision of project finance on international construction firms. The clear implication of this, in what is a buyer's market for construction services, is that international contractors are increasingly having to compete upon the attractiveness of the financial package they offer rather than the technical services they provide. This form of competition has become the standard competitive practice in the highly indebted LDCs.

The importance of financial competition within the contemporary international construction industry cannot be overstressed. A merchant banker recently suggested that within the international construction industry:

> What we now face is a buyer's market and probably the major factor in awarding contracts is the financing package with which the different companies or consortia bidding for a project can come up with. (Financial Times, 5/3/84)

The Financing of International Projects

Contractors in the survey expressed similar feelings. 90% of the survey participants suggested that the most important factor to have emerged in the industry over the past 5-10 years is that the contractor has been increasingly relied upon to provide finance for construction projects. Other empirical evidence tends to back this up - 50% of the top 250 international contractors in the ENR annual survey for example reported that they were asked to provide 'financial engineering' (2) with their bids in 1983 (ENR, 2/8/84).

In the next section we introduce a brief outline of sources of finance open to MNEs, and how these relate to international contractors and the provision of project finance. This is followed by a discussion of how financial competition takes place within the international construction industry.

7.1 FINANCING OPERATIONS IN THE MNE

It is inevitable that in the lifetime of the MNE funds will be required for overseas expansion either through the establishment of a new subsidiary or distribution outlet, or the enlargement of existing facilities in an overseas market. Eiteman and Stonehill (1982) suggest that these funds may come from three possible sources:

1. Funds generated internally within the foreign affiliate.
2. Funds generated internally within the parent company.
3. Funds from sources external to the corporate family.

The nature of internal and external funding are now discussed in more depth.

7.1.1 Internal funding

Kim (1983) suggests that internal funding can be subdivided into three possible sources:
 1. Funds provided from subsidiary operations. Once the subsidiary has the capacity for earning a return in an overseas market retained earnings and depreciation are the dominant source of funds. Restriction on repatriation of funds may be the major impetus behind the reinvestment of income of foreign subsidiaries in overseas markets.
 2. Funds from the parent. The three major types of

funds supplied by the parent are equity contributions, direct loans, and indirect loans under parent company guarantees. Reasons for this form of financing generally revolve around financial benefits to the corporate family that result from this policy. Robbins and Stobaugh (1974) suggest that parent funding of foreign subsidiaries may be a major source of financing for three reasons:

a. Loans can be used to repatriate funds to the parent where host country restrictions would normally prevent repatriation on such a scale.
b. Loans can be converted to equity relatively easily so that the parent can determine the equity balance without host government objections.
c. Internal loans may be used to reduce taxes because of different fiscal policies in the home and host country.

All three may be considered potential transactional advantages of parent funding in terms of the eclectic framework.

3. Loans from sister subsidiaries. The availability of credit from other subsidiaries in addition to parent company funds vastly expands the number of possibilities of internal financing. However, exchange controls and restrictions on capital movements imposed by many countries may limit the range of possibilities for intersubsidiary loans.

7.1.2 External sources of funds

External sources of funds will normally be required where the MNE needs more funds than the amount that can be reasonably generated within the corporate family. External funds can come from three potential sources:

1. Sources in the parent country, eg banks and other financial institutions, securities markets or money markets, and home government assistance (where available).
2. Sources in the host country, eg local currency debt, joint business ventures with local owners, host government incentives (where available).
3. International sources, eg Eurocurrency debt, third country debt, development bank financing.

Given the nature of internal and external debt, Eiteman and Stonehill (1982) suggest that choice among the source of funds should ideally involve simultaneously minimising the cost of external funds after adjusting for foreign exchange

The Financing of International Projects

and political risk, and choosing internal sources in order to minimise world taxes and political risk. In practice however these aims are likely to be antagonistic so that the tendency is to place more emphasis on one of the variables at the expense of the other. Ultimately this will depend upon the proportion of the debt financed internally and externally, which will be influenced by the internal availability of capital and corporate aims and policies concerning the use of internal and external sources of finance.

To analyse the determinants of choice of funding within the international construction industry, it is necessary initially to point out that the contractor requires two types of finance for overseas operations:

1. Working capital. The international contractor will require working capital to maintain production and production facilities (ie human capital and subsidiary operations) and expand overseas via increasing facilities and bidding for projects. In the process of funding working capital the contractor's choice of financial source will be similar to that of the multinational manufacturer, as the decision on whether to use internal or external funding will come down to the internal aims of the company and direction of its financial policy. Generally, internal capital will be preferred to external funding for working capital, because this enables the contractor to coordinate resources and thereby gain transactional advantages as discussed above.

2. Project finance. In the case of project finance, which forms the basis of financial competition, the contractor faces a different situation to the provision of working capital. Because of the nature of the final product project finance is required by the client in the pre-construction and construction periods to finance the construction of the final product. The provision of project finance by the contractor is therefore likely to be heavily influenced by the cost and availability of capital, and may be regarded as a form of internalisation by the contractor of part of the vertical process of construction activity to give the contractor a more diverse product differentiation strategy (as discussed later in this chapter and in Chapter 8). This clearly bestows transactional advantages on the contractor which are not available where the firm does not provide project funding.

The provision of project finance though not unique to international construction, may be regarded as a consequence of the production process of construction. Because demand necessarily follows supply within the industry, effective demand requires that finance for construction is necessary in the pre-supply phase. This will be required not only to pay for the services and materials needed for construction over the production process, but will also be demanded by the contractor to prevent default of payment. If default were to occur on completion of the project the nature of the contractor's product means that it could not be withdrawn from the final product and as such is not resaleable. To prevent this potential loss, the contractor will therefore require pre-payment or continual payment for services throughout the life of the project.

Where the client cannot obtain finance for construction the onus will fall upon the contractor to create effective demand for construction works. Where this is necessary, we argue that project finance will be arranged by the contractor from external rather than internal sources of the company for two reasons:

1. The capital is not exploited for internal use by the firm and therefore the company is unlikely to approve of using internal funds for project finance when it could gainfully be employed elsewhere as working capital of the firm to increase internal coordination, and therefore enable the company to exploit transactional advantages open to the contractor. Additionally because the finance would be externalised outside the company, potential problems to the contracting company's internal organisation which would be caused by loss of the funds (ie borrower default) are too great for internal project financing to be considered.

2. By its nature, project financing in international construction involves millions and sometimes billions of dollars of capital which would not be available internally within most international construction firms, and would not be offered where available because of the high exposure to risk the contractor would face by using internal funds for project finance, as mentioned in 1 above.

Because of the size of the capital involved in project finance in international construction and the high level of potential risk the contractor faces by obtaining project finance on the external market, a significant feature of the financing of projects in international construction which separates international contracting from many other multinational industries is the increasingly prominent role of home governments in providing means for obtaining finance for this purpose in the form of export credit and bilateral aid made available to the contractor. This policy is thought to reflect two factors:

1. The need for finance beyond most contractors' means for project finance in developing regions, and the high risk exposure that characterises project finance in international construction.

2. Construction forms part of a greater policy by many governments to promote exports via financing agreements.

Additionally however the increased role of governments to finance this type of operation has implications for the nature of country specific O advantages within international construction. Because of the size and ability of the home government to provide resources above that of any enterprise, the contractor with the highest level of government support in this respect is likely to be at the greatest advantage where project finance is required. In this chapter we argue that this is more likely to be affected by the attitude of the home government rather than institutional or capital market imperfections that may affect the cost of capital (ie Aliber 1970, 1971), because the home government will have access to both funds and information beyond that of any institutional investor, which will provide the basis for the financing. The objective here therefore is to outline the policies of home country governments in this respect, and illustrate how these influence country specific competition within the international construction industry.

7.2 THE EXPORT CREDIT MECHANISM

The term 'export credit' refers to the set of facilities that the home government makes available to exporting enterprises so that the exporter faces minimal risk exposure and financial obstacles as a consequence of exporting. As such, export credit is a mixture of insurance and banking mechanisms which help exporters overcome the major problems associated with overseas trade so that these exporters may enter markets that otherwise would be extremely costly to enter with only commercial support in the form of bank loans and insurance. Export credit may therefore be considered an aspect of home government support for overseas trade.

All major exporting nations have an export credit system to protect exporters, and though no two are alike because of the different political and economic systems in which they have evolved, the problems faced by exporters are common to all countries so that they have developed within similar parameters and have faced regulation by the same international guidelines. The basic services are therefore likely to have common elements, and the facilities provided by most export credit agencies fall into two categories, insurance and financing.

7.2.1 Export credit insurance

The insurance aspects of export credit have been developed by the major exporting nations to reduce the risks of exporting, most notably the risk of non-payment that can be caused either by political or by commercial circumstances. Both are generally covered by export credit insurance. Within most systems the contractor can cover these risks either by a 'global' or a 'specific' policy. A global policy involves the exporter nominating all the firm's export orders for cover, within the parameters of that policy. A specific policy will be used when a contract is non-repetitive, the credit period substantial and the operation large in terms of financial resources. It is therefore likely that specific insurance will be used by the international contractor more than global insurance. In both types of policy the contractor will pay a premium for the use of the service (related to the level of risk and the size of the contract), and will share a proportion of the risk though this is not likely to be more than 10-15% of contract value.

The availability of export credit insurance has several direct benefits for the contractor:

1. Export credit insurance encourages the exporter to offer competitive terms and lessens the effects of exogenous variables so that the exporter may penetrate risky foreign markets. In international contracting this is obviously of great benefit.

2. Export credit insurance protects the company's foreign returns (either potential or actual), which is possibly the riskiest part of the firm's asset portfolio.

3. Export credit insurance gives the contractor greater flexibility and financial liquidity by ensuring future payments and therefore permits the firm to undertake

company objectives on the basis of these returns.

Given that export credit insurance schemes vary from country to country, the insurance aspect of export credit may provide a potential country specific O advantage as discussed later in this chapter.

7.2.2 Export credit financing

Finance in export credit involves the export credit agency supplying the means to either medium or long term credit for payment of the goods and services in a contract. This prevents the exporter facing a financial 'bottleneck' of funds, during or after contract completion, which may be caused by the client wishing to pay for the goods/services over an extended period of time. Finance for export credit can be made available as supplier or buyer credit and does not necessarily imply that the export credit agency actually supplies the finance.

A supplier credit is an arrangement under which the client is allowed deferred payment on the good/service, but the supplier receives payment as soon as possible after the completion of the contract and therefore effectively provides credit through the supplier to the client. This mechanism involves the buyer being obliged to make extended payments to the supplier through a stream of promissory notes or bills of exchange. Once received by the supplier, these notes are bought or discounted by the supplier's bank on arrangement so that the bank receives the extended payments from the buyer in the future, and the supplier receives immediate payment. Throughout this the export credit agency guarantees the notes against default by the buyer, so that the bank who accepts the notes takes no risk. The credit agency therefore effectively shoulders the risk of default of the buyer, providing all goods and services included in the transaction are carried out to a satisfactory level.

Under a buyer credit, the export credit agency insures the loaner of the capital (which may be either the export agency itself or some commercial institution) against default where a specific loan is arranged between the lender and the buyer or some borrower in the buyer's country connected with the buyer. The buyer pays the loan directly to the lender, and the supplier of the good/service is paid by the lender on delivery of documents suggesting that the

The Financing of International Projects

supplier has completed the contract to the initial agreement of the supplier and buyer. The export credit agency, as in the supplier credit, ensures that the lender receives payment and likewise insures the supplier.

An extension of the buyer credit is a line of credit, which is a single loan agreement set up to finance a number of separate contracts. Additionally a project line of credit is where there are a number of contracts for one project buyer financed by export credit funding. In both cases the finance is acquired before a borrower is found, and then advertised for a suitable client. This will require the export credit agency's approval if the agency is to endorse the financial transaction as official export credit.

For the construction contractor the buyer credit is more commonly used than the supplier credit because it enables the contractor to receive advance payments for services and materials throughout the life of the project, and limits the recourse and exposure to the construction firm. The analysis of export finance here is therefore restricted to buyer credit, to concentrate the analysis on only those aspects of export credit which affect international construction.

The funding of export credit will reflect the structure of the export credit agency. In some countries for example the agency will be both an export bank that will provide finance for export credit and underwrite the insurance risks (eg Export-Import Bank of America), while other countries' agencies provide insurance only, although they clearly have links with potential sources of export credit funding (eg Export Credit Guarantee Department of the UK). For the export agency with facilities for insurance only, the source of funds may be the domestic, international, or even the buyer's domestic capital market depending upon the financial arrangement, while the export agency which is also an export bank will generally depend upon government expenditure allocation as a source of finance. In practice both will have advantages and disadvantages that may be the source of country specific O advantages in the field of export credit.

7.3 EXPORT CREDIT FINANCING AS A COUNTRY SPECIFIC O ADVANTAGE

The attractiveness of export finance as a country specific O

The Financing of International Projects

advantage in a situation characterised by financial competition is likely to come down to two issues:

1. The level of services the export agency provides. The greater the number of facilities connected with export credit the agency provides, the more contractors from that country will benefit from greater government support than lesser backed contractors.

2. The cost of the export finance to the buyer. Cost of the finance is obviously likely to be the major determinant of success in a bid where financial competition ensues and two or more contractors are able to provide the capital necessary. Financial competition can therefore be considered a mixture of product differentiation and price competition for the international contractor.

These are now discussed in turn.

7.3.1 Export credit services available to the contractor

It is clear from Chapter 5 that the level of export credit services provided will depend upon the home government's attitude to intervention, so that the amount of support is likely to vary from country to country. Because of the scale of credit and insurance required for international construction, government support can rarely be replaced by a service in the market providing similar services so that the more government support the contractor receives as official export credit the more competitive that contractor is likely to be relatively, since this will both allow the contractor to cover risks better (and therefore penetrate more foreign markets), and also to compete on financial terms with a wider range of services.

Table 7.1 adopts the same approach as Table 5.16 for general government support to contractors, though in Table 7.1 the factors relate to aspects of export credit only. Turkey is not included in the table because the Turkish government does not provide an export credit system.

It is clear from Table 7.1 that France has the most comprehensive export credit financing system, which together with Table 5.16 suggests that the level of French government involvement in overseas operations of home country contractors is high relative to most of the other major contracting countries. Japan also has an extensive export credit system which is consistent with the level of Japanese government support as illustrated in Table 5.16.

Table 7.1: A Comparison of Export Financing Systems of Major Contracting Countries

	France	Germany	Italy	Japan	UK	US	Korea
Direct supplier credits	Yes+	Yes+	Yes+	Yes+	Yes+	Yes+	Yes+
Direct buyer credits	Yes+	Yes+	Yes+	Yes+	Yes+	Yes+	Yes+
Refinancing facilities	Yes+	Limited-	Rare-	Yes+	Yes+	Limited-	Limited-
Mixed credits	Yes+	Yes+	Limited-	Yes+	Limited*	Limited*	Rare-
Export insurance	Yes+	Yes+	Yes+	Yes+	Yes+	Yes+	Yes+
Export loan guarantees	Yes+	No-	No-	Yes+	Yes+	Yes+	No-
Exchange risk insurance	Yes+	Yes+	Yes+	Yes+	Yes+	No-	No-
Inflation risk insurance	Yes+	No-	No-	No-	Yes+	No-	No-
Bond insurance	Yes+	Yes+	Yes+	Yes+	Yes+	No-	Yes+
Ceiling on authorisation	No+	Yes-	Yes-	Yes-	Yes-	Yes-	Yes-
Bidding cost insurance	No-	No-	No-	Yes+	No-	No-	Yes+

* Official aid component of export credit only to match competitors' bids where necessary.

Source: Dunn and Knight (1982), OECD (1976), Demacopoulos and Moavenzadeh (1985)

The Financing of International Projects

Although the Italian system looks fairly competitive in Table 7.1, budget constraints limit the effectiveness of the system; a similar argument can be put forward as an explanation of the limited services the Korean government provides. Germany (3) and Britain have fairly comprehensive export credit systems while Americans receive the least amount of official support in this respect.

To analyse the components of Table 7.1 further, the table can be split up into those aspects of export credit financing and insurance that provide the most support for international contractors:

1. Financing. Given the nature of export credit financing, clearly the most important feature of any system is the provision of supplier and buyer credits, though in the case of construction buyer credits are especially relevant. All the countries in the table supply both buyer and supplier credits, suggesting that in most cases these provide the basis of export credit finance. The ability to refinance a credit and offer mixed credits is not so universal. Refinancing enables the client to raise funds to pay back the original loan from sources other than the export bank, while mixed credit involves the provision within the credit for a proportion of public funds (usually in the form of aid) that lowers the effective interest rate on the credit. Both may be useful where financial competition ensues.

Most of the countries in the table have some form of refinancing facility though for German, Italian, American and Korean exporters this service is limited to specialist cases so that refinancing is not a common feature in these systems. France, Japan and Britain have comprehensive refinancing arrangements. The provision of mixed credit is something that is dealt with in more detail in the following section, but it is interesting to note that this forms the basis of financial competition since it allows the contractor to compete on the interest rate of the credit given to the buyer. France and Japan are the major proponents of mixed credits though Germany, Britain and America will back a mixed credit agreement by providing part of the credit as aid, (although in the UK and US systems this is only usually likely as a means of matching competitors' credit terms). Italy and Korea do offer mixed credit but are constrained by the amount of finance they can afford for such a system.

2. Insurance. There are various forms of insurance open to international contractors, within export credit systems which are likely to enable the contractor to cover

risks of entering a market and minimise potential losses caused by political or commercial circumstances. The provision of export insurance usually covers these basic risks, and as with supplier and buyer credits forms the basis of most export credit systems. As such export insurance is provided by all the government systems in Table 7.1.

Other forms of insurance in the table are export loan guarantee (to insure an export loan against default of payment), exchange rate insurance (to insure the exporter against loss of returns on conversion to the domestic currency), inflation risk insurance (to protect the exporter from the effects of escalation of costs), bond insurance (to prevent loss through unfair calling of bonds), and bidding cost insurance (whereby the contractor receives part of the costs of bidding if the tender is unsuccessful). The 'ceiling of authorisation' in Table 7.1 refers to whether the export system is limited to insuring only up to a specified amount. Where this stipulation is waived the insurance system as a whole benefits.

Clearly where these types of insurance are available the contractor will benefit from lessening risk exposure, but also from lowering the overall cost of overseas operations that may be reflected in bidding prices. French contractors seem to particularly benefit from extensive export insurance cover, and to a lesser extent the UK also has a comprehensive insurance system. Germany, Japan, and America have systems that are not as competitive, but supply several types of insurance which are beneficial to international contractors from these countries. Italy's insurance system is less comprehensive while the Korean government support in this respect is fairly basic, although the provision of bidding cost insurance is likely to be a major benefit to Korean contractors.

Table 7.2 summarises the percentage cover that the major contracting countries give on commercial and political insurance (ie 'export insurance' in Table 7.1) to illustrate the relative competitiveness of the major contractors' insurance schemes. The table illustrates that insurance cover is likely to be similar for all the countries, although Japanese and especially Italian contractors may be at a disadvantage where minimum cover is offered. Data on the level of Korean cover was not available.

It is clear from the discussion above that export credit finance systems vary from country to country, and that the facilities provided by the home country export credit agency

Table 7.2: Percentage Cover of Commercial and Political Risks of Major Contracting Countries

Country	Commercial Risks+	Political Risks
France	90-95	90-95
Germany	95	95
Italy	10-85	90
Japan	60-90	90
Britain	90-95	90-95
America	70-90	70-90

+ Includes insolvency of the buyer and failure to pay local currency deposit at the due date.

Source: Demacopoulos and Moavenzadeh, 1985.

are likely to be instrumental in the success of contractors abroad, particularly in politically risky areas or where financial competition ensues. This is broadly in line with Dunning's classification of a country specific O advantage, and suggests that some contractors may have an inherent advantage resulting from the level of support the home government is prepared to give in this area. The analysis clearly illustrates that there are discrepancies between the support contractors get, which may be relevant to the ability of contractors to enter new and risky markets in search of project work. Since the majority of work for international contractors is likely to be in these areas, this may be a significant feature of competition within the industry.

7.3.2 The export credit 'war'

Although the provision of export credit services differs between countries, there are international regulations concerning the use of export credit aimed at preventing competitive subsidisation of export finance to limit competition based upon the attractiveness of the financial terms offered rather than characteristics of the good/service purchased. Foremost amongst these accords is the 'Arrangement on Guidelines for Officially Supported Export Credits' established in 1978 by the OECD and generally known as the 'Consensus'. The guidelines take the form of a matrix which shows the minimum interest rate and the maximum repayment period an export credit agency can offer according to the status of the buying country. The matrix is reproduced in Table 7.3, and shows for example that for an intermediate country buyer the export credit agency may offer export credit for up to 8.5 years and can charge a minimum interest rate of 11.2% on a credit exceeding 5 years. In the matrix, relatively rich countries are those with GNP per capita greater than $4,000, intermediate countries are those with GNP per capita between $681-$4,000, while relatively poor countries are those with GNP per capita of less than $681.

Table 7.3: OECD Arrangement Terms on Officially Supported Export Credits 1985

Importing Country	Maximum Period for Repayment		
	2-5 years	5-8½ years	8½-10 years
	(Annual interest rate in %)		
Relatively Rich	12.00	12.25	-
Intermediate	10.70	11.20	-
Relatively Poor	9.85	9.85	9.85

Source: Finance and Development, September 1985

From its inception the Consensus has had problems because of the diverse opinions of the member countries on the function of export credit, and it is this that has paved the way for financial competition. Although the Consensus has aimed to prevent competition on financial terms (which it may have done to an extent), the single rate of interest has led to deviation from that rate by certain countries as an incentive to the buyer to purchase a specific good/service on the basis of the cheaper finance made available by the export credit agency. This is the state of financial competition that has developed in the international construction industry. The lack of development finance and the buyer's market in the industry has increasingly led to the situation where contractors are competing on the conditions of the financial package they offer more than any other aspect. Hence the attitude of the domestic export credit agency towards providing competitive finance may be considered an essential element of the competitiveness of that nationality's contractors where financial competition ensues. Competitive finance in this context may take two forms.

7.3.2.1 Market rate subsidy
The initial Consensus interest rates were set at a time when market rates were comparatively close to one another. As a result, once the Consensus was introduced all those participants were giving roughly the same amount of subsidy on official export credit (ie the difference between the market rate and the Consensus rate). In the period following this however market rates have diverged substantially with the result that in the past, several agencies have had to increase the level of export subsidies dramatically to promote export credits at Consensus rates because higher interest costs have had to be paid to obtain funds in the market to sustain official credits. In 1982 for example government bond rates averaged 15.6% in France and 12.9% in Britain while the minimum allowable rate specified in the arrangement for export credit to developing countries was 10%. Given this situation, objections about the level of the Consensus rate and permissible subsidies have reflected nationalistic views. The Americans for example have tended to argue for high rates because EXIMbank (the export bank of the United States) has limited funds for interest rate subsidy (because of government reluctance to provide subsidies) and is a major exporter of large capital goods which require long payback (and therefore high overall interest rates), while France at the other extreme as a country that relies heavily upon international trade sees interest rate subsidy as a valid means of promoting trade and therefore in the past has set aside a large proportion of government finance for this purpose.

Although several countries' market rates have increased above the arrangement rate, a few have experienced interest rates below the matrix guidelines. As an option to being penalised for this, West Germany introduced the concept of 'pure cover', whereby exporters from that low interest rate country (LIRC) offer commercial rates with official involvement confined to a guarantee on the credit. Where this has occurred, exporters from that LIRC are clearly at an advantage to those offering Consensus rates. For this reason several major exporting countries have objected to pure cover, including France and America, although the practice has now been generally accepted.

As a consequence of the divergences in the subsidy

levels and market rates, the OECD arrangement was modified in 1983. The modifications included the adoption of a formula that adjusts minimum rates regularly and automatically in line with market rates, and also the adoption of individual formulae for currencies where market related 'commercial interest reference rates' serve as the minimum allowable official lending rates that may be used by any member export agency wishing to undertake credit in that currency.

Linking the arrangement minimums to market rates has in principle eroded direct subsidies on interest rates so that from 1983 onwards this form of subsidised export credit has not been common. Minimum rates are not always observed however, and the level of subsidisation still grows when market rates diverge from the biannual adjustments of minimum rates, though the impact of this has been reduced considerably by the arrangement modifications.

7.3.2.2 Mixed credit financing

Although interest rates have been given the most attention in terms of the OECD arrangement, mixed credit is possibly the more common form of financial competition and the major threat to the arrangement. Mixed credit involves the mixing of conventional export credit with aid funds to offer a composite interest below the matrix rate. The theory behind this can be illustrated using the formula given by Demacopoulos and Moavenzadeh (1985) which shows the net present value (NPV) of cash flow of finance:

$$A = P - \sum_{t=1}^{V} \frac{P.z}{(1+R)^t} - \frac{P}{(1+R)^V} \tag{1}$$

Where
P = principle amount of export credit finance
V = number of years P is repaid after
z = concessional interest rate
R = Consensus interest rate
A = level of aid in the credit.

To illustrate the effect on interest rates if the proportion of aid in the loan is increased, equation (1) must be manipulated to isolate z. From (1):

$$\sum_{t=1}^{V} \frac{P.z}{(1+R)^t} = P - \frac{P}{(1+R)^V} - A \tag{2}$$

The Financing of International Projects

If:
$$\sum_{t=1}^{v} \frac{P.z}{(1+R)^t} = z \cdot \sum_{t=1}^{v} \frac{P}{(1+R)^t}$$

then
$$z \cdot \sum_{t=1}^{v} \frac{P}{(1+R)^t} = P - \frac{P}{(1+R)^v} - A \qquad (3)$$

which gives
$$z = \frac{P - \frac{P}{(1+R)^v} - A}{\sum_{t=1}^{v} \frac{P}{(1+R)^t}} \qquad (4)$$

From equation (4), we can illustrate the effects of aid in a mixed credit. We assume here for this empirical exercise that the export credit is for $20 million, at a consensus rate of 10% with a payback period of 20 years. Table 7.4 illustrates that in this case, where the aid component of the credit is zero the effective interest rate is that given by the consensus, although beyond this as the proportion of aid in the export finance rises, the interest rate falls so that the credit will become cheaper in competition with credit at the consensus level. The exercise therefore illustrates that mixed credit may be an effective way of competing where cheap finance is required.

Table 7.4: Effective Interest Rate with Aid Component of Export Finance

% of export credit as aid	Concessionary interest rate
0	10.00*
10	8.83
20	7.66
30	6.48
40	5.31
50	4.13
60	2.95
70	1.78
80	0.60
90	0.57
100	0.01

* i.e. rate of interest same as Consensus

The provision of soft loans (ie loans that have interest rates below the market level and/or longer payback periods) as mixed credit can generally take two forms:
1. Tied aid credits. These are broadly defined as credit for development purposes and are offered by the donor country on condition that all goods and services required for development with that finance come from the donor country. Tied aid may be either project finance or a line of credit. Untied aid refers to funding that can be used to purchase goods and services at least from all OECD countries and developing countries for development.
2. Associated financing transactions. This is a more narrowly defined concept than tied aid, combining two or more of the following:
   a. Overseas Development Assistance (ODA) with a grant element of at least 25%;
   b. other official flows (except as in item c) within a grant element of at least 20%;
   c. officially supported export credits, or other funds at or near market terms.

The provision of tied aid within export credit is a controversial issue within the OECD because of the conflict of views on the use of aid by member countries, but it has been catered for within the 1983 modifications to the arrangement. Under the amended arrangement aid funds may be mixed with export credit only when the overall grant element is at least 25% (to lessen commercial considerations of including aid). In practice the use of more concessional tied aid (eg with a grant element exceeding 25% of the credit) has become more common in export credit financing as a means of competition, because it is not restricted by the arrangement.

The offering of aid as part of an export credit package generally involves some form of official mechanism within the government that enables the exporter to receive a proportion of the credit as aid in the form of an interest free loan, a gift, or a credit with a low interest rate and/or long payback period. In the UK this government mechanism is the Aid and Trade Provision (ATP). Founded in 1977, the ATP was set up as a defensive measure against 'predatory financing' by competitor countries. ATP accounts for approximately 5% of the UK total aid budget, and generally aims to provide funds in countries not normally covered by the UK's aid programmes, or where the aid allocation has already been committed. ATP is offered direct to public authorities on condition that the relevant contract is awarded to a British company, although there are provisions where a non-UK element may be included if the foreign goods or services are vital to the project. In the US, the Agency for International Development (AID) provides a similar mechanism to ATP in the form of the Trade Development Programme (TDP). In addition, AID provides dollar loans outside TDP for the purchase of US goods and services, though for the most part these are partially untied (ie the difference between tied and untied aid).

In both the ATP and TDP mechanisms there are conditions that must be fulfilled before the system can be incorporated into a conventional export credit package. The most important of these is that there is evidence of foreign competition officially supported by a measure of aid or other generous terms that contrast with the Consensus, so that neither the UK nor US will generally offer mixed credit in the form of tied aid as an initial incentive to prospective buyers. In both cases the provision of aid is used as a means of matching competition, which in theory enables the contract to be won on non-financial terms. Other conditions of ATP include that the aid should help create industry and jobs abroad without over capacity, the technology involved should be appropriate to the recipient country, and that there should be substantial follow-on orders on normal commercial terms.

The basis of financial competition on mixed credit, which is significant to the discussion of financial competition within the international construction industry, stems from the fact that different countries view tied aid differently according to industry and government policies and objectives. Because mixed credit is generally tied, then where the home government approves of mixed credit contractors from that country will benefit over other nationality contractors who do not receive government support in this area, since finance on the open market is unlikely to be obtainable at the same competitive rate. It is therefore probable that the government provision of finance for tied aid will be the most instrumental country specific O advantage in financial competition.

Of the major contracting countries, French contractors are likely to benefit the most from tied aid finance because of the French government's attitude in this respect. The French government, as the initiator of mixed credit ('credit mixte') has continually argued for mixed credit on the basis that this provides a channel for development assistance which has a marginal effect on exports. However, this policy is also likely to reflect the fact that the French consider the export sector of primary importance to the economy, that it 'deserves unlimited government assistance'; Dunn and Knight (1982) estimate that approximately 10% of all official credit transactions in France are financed through mixed credit.

In the wake of this French policy, other contractors' governments have been increasingly drawn into providing competitive finance to enable domestic exporters to compete abroad. In a recent survey by the Hawker Siddeley Group (1985) it was argued that West Germany, Japan, Canada, and to a lesser extent America, all provide extensive generous and flexible facilities for offering soft loans and tied aid in support of exporters' tenders. While the Hawker Siddeley analysis was confined to railway contracts, this situation is reflected in the international construction industry. The provision of soft loans by the American and Japanese governments in China for example have been instrumental in contractors from these countries receiving major construction contracts in the country, while French financial assistance to Algeria and Francophone Africa has helped in maintaining French contractors' workloads in Africa in the past.

As a country specific O advantage within the eclectic framework, the relative competitiveness of contractors in financial competition is likely to depend upon the level of financial support the contractor's government gives for tied aid, and the nature of the facilities under which aid finance is made available. It is argued here that in both cases UK contractors may be at a disadvantage with respect to competitors, and therefore we now discuss each of these factors in turn.

    1.   Level of financial support. Generally the greater the home country's commitment to bilateral rather than multilateral aid, the more this form of financial support can be used for commercial purposes along the lines of financial competition as discussed above. This reflects the fact that in multilateral aid investment decisions are usually taken by international quangos so that the gearing (ratio of return to initial contribution) is often unpredictable and frequently low, while in bilateral aid a more commercial emphasis may be placed upon the use of the finance.

Table 7.5 illustrates the aid commitment of the major contracting countries for 1983; Korea and Turkey are not

Table 7.5: Aid Commitment of Major Contracting Countries, 1983

	Bilateral		US $ Million Multi-lateral	Total	Ratio of Bilateral/Multilateral
	Total	of which grants			
France	3809.5	3151.4	669.8	4479.3	6:1
Germany	2363.7	1233.1	1081.2	3444.9	2:1
Italy	–	–	–	–	–
Japan	2809.1	920.3	1340.2	4149.3	2:1
U.K.	929.0	873.2	746.7	1675.7	1:1
U.S.	5704.0	4142.0	2503.0	8207.0	2:1

Source: OECD 'Development Cooperation 1984 Review'

included because they are net recipients rather than net providers of aid. The table shows a marked difference between the countries both in the level of aid provided and the proportion of this finance that is given as multilateral or bilateral aid. Given the export promotion policies of the French government, it is perhaps not surprising that the ratio of bilateral to multilateral aid is overwhelmingly in favour of bilateral aid at 6:1, and that the French provide the second largest aid commitment in the table. American, German, and Japanese contractors are also likely to benefit from a generous provision of bilateral aid from their home governments, as shown by the total amount of aid provided and the ratio in the table. As a contrast, Italians are likely to be at a distinct disadvantage to these countries since in 1983 the government provided no financial aid either bilaterally or multilaterally.

One of the major elements of Table 7.5 is that UK contractors appear to lose out with respect to competitors in terms of aid, as illustrated by two points within the table:

a. Total finance made available for aid by the UK government in 1983 was half that of Germany's aid commitment, the second lowest aid contribution in the table.

b. The ratio of bilateral to multilateral aid in the UK case was 1:1, which is lower than that of major competitors in the table even when the lower UK total aid figure is not taken into account.

Table 7.5 therefore suggests that the scope for UK contractors to undertake financial competition on the basis of mixed credit is likely to be limited. This argument is further justified in Table 7.6 which shows the tying status of bilateral aid. Clearly where this is partially tied or tied then there is scope for financial competition. The table illustrates that UK and German contractors may be at a disadvantage to American, (4) French, and Japanese contractors who are likely to have access to more extensive funds for tied credits. This relative advantage/disadvantage may be considered an important country specific O advantage where financial competition is at a premium.

2. The nature of credit facilities. Although the UK export credit system may be considered comprehensive in relation to competitors' export credit agencies, in the context of financial competition the ATP programme in particular is classed by many as the major instrument available to contractors to counter national competition on

Table 7.6: Tying Status of Bilateral Aid Provided by Major Contracting Countries 1983

(US $ Million)

	Bilateral Aid	
	Untied	Partially Tied/Tied
France	1365.0	2444.5
Germany	1659.7	704.0
Japan	1561.4	1247.7
UK	238.4	690.6
US	2210.0	3494.0

Source: Ibid.

financial terms. However within the UK there have been several criticisms of how the ATP and bilateral aid programmes are run, which claim that UK exporters (including contractors) are losing out in world markets because these mechanisms are not being used efficiently. Three reasons are given here for this:

a. Because of the divergence in interests between the UK Treasury (who view ATP and tied bilateral aid as a subsidy) and the Department of Trade and Industry (who see these mechanisms as a legitimate way of promoting trade), the amount of the aid budget set aside for ATP is only approximately 5% of the total (ATP in 1983/84 was set at £66 million). The Overseas Project Board (5) argue that ATP is an efficient tool of export support and should take a greater share of the bilateral aid budget, which is currently more developmental than commercial.

b. Whereas other governments offer 'pre-emptive' aid, the UK government will only match, not initiate financial competition, and this is only possible where the contractor can provide evidence of competitors carrying out such competition. Since evidence may take time to accumulate, and the contractor who gets ATP is only matching a deal already 'on the table' from another contractor, these conditions are likely to be highly restrictive in practice.

c. Generally where evidence can be found of financial competition, the UK will only match by offering aid plus separate finance at the Consensus rate of export credit. Competitors however are for the most part offering a mixed credit package where the aid is used to cut the overall rate and extend the term of the loan beyond that permissible within the OECD arrangement; in the Hawker Siddeley Group report (1985) for example the authors quote a West German mixed credit package for Turkey repayable over 30 years with a 10 year grace period, compared with a UK offer of fixed aid and credit at 11.35% over 10 years.

7.3.3 Alternatives to the export credit 'war'

The situation presented in the previous section suggests not only that financial competition based upon export credit and aid financing is of major importance to the international construction industry, but also that certain countries may be losing out in world markets because of domestic restrictions to competition in this sphere (notably the UK and Italy) while others' success in certain regional markets may be attributable to their financial competitiveness in this respect (especially France and Japan).

Although several alternatives to the 'war' have been discussed by the major exporting nations, the impact of these so far has been minimal. This is reflected in the fact that the American government (a major critic of subsidised export credit) has recently sanctioned a $266m 'warchest' to assist US companies compete abroad on the basis of mixed credit.

It is suggested here that there are two major reasons why subsidised export credit is likely to remain a problem for some time:

1. The system remains too open ended so that member countries may diverge from the arrangement whenever it suits them, without penalty or recourse. Pearce (1980) for example cites the case of Japan offering credit to China at 6.5% over 5 years (considerably below the OECD Consensus at the time) by describing the credit as import financing, a category not covered by the OECD arrangement. (6) However, the most common form of divergence is that of mixed credit, as discussed above. Where the system is ambiguous and non-restrictive then there is clearly incentive for opportunistic gains where

The Financing of International Projects

possible.

2. The market situation. Because international construction is likely to be a market characterised by a high level of sellers and a low incidence of buyers for some time, there is incentive to undertake financial competition to obtain work where project finance is necessary. It is unlikely that restrictions placed upon subsidised export credit will be effective where this situation prevails since the seller has the motivation to go against the agreement to capture a larger market share. This is especially likely because France will in all probability maintain a policy of competing on export credit terms regardless of the arrangement, so that competitors must also adopt a similar policy.

The two reasons when taken together suggest a circular situation where financial competition may ensue because of the nature of the arrangements, but the arrangements cannot be changed to effect because financial competition ensues. One possible way to break this circle would be to introduce one or more of the multilateral credit agencies as a 'clearing house' of export credit to developing countries so that competition could be monitored and sorted on non-financial terms, although this would face objections from those countries that argue for the use of financial competition such as France and Japan.

Given the market situation and the attitude of some of the major contracting countries towards the use of mixed credit, it is likely that financial competition may remain a significant feature of the international construction industry for several years to come. As long as this situation is prevalent financial competition is likely to provide a major element of country specific O advantages within the international construction industry which generally will be manifest in export credit finance and the provision of cheap capital for mixed credit arrangements. (7)

7.4 SUMMARY

From the discussion of project financing in international construction above it is clear that international contractors face a relatively unique situation regarding financial competition within the industry, which is characterised by three rather paradoxically related factors.

1. A high level of official involvement by all the major contracting countries' home governments.

2. International guidelines that aim to replace the market for project capital with stabilised and uniform rates to prevent competition.
3. Based upon and affected by 1 and 2, the emergence of a form of competition that the contractor is powerless to change or exploit without extensive government help.

In this sense economic theory of international finance such as the work of Aliber (1970, 1971) cannot be said to have any direct relevance to the situation, though clearly more macro oriented political and economic analysis of the resources and attitudes of home governments is likely to be useful to any country specific analysis of the phenomenon. Such detailed analysis is beyond the scope of this research, even though it is likely to be relevant to the discussion of the financing of overseas operations of MNEs where some form of project finance is required.

Although we argue that international financial theory is not of direct relevance to this discussion of project finance, international production theory in general and the notion of competition using O advantages in particular offers an indication of how financial competition has emerged, and also suggests that this form of competition is likely to remain important in the industry as long as there are benefits to the contractor and the home government to compete in this way. As such the theory of international production appears most relevant as an explanation of the predominance of financial competition within the international construction industry.

7.5 FOOTNOTES

1. I would like to thank Patrick Hodgson of the Major Projects Association (Templeton College, Oxford), for helpful advice on the nature of export finance and comments on an earlier version of this chapter.
2. The term 'financial engineering' refers to the demand prospective buyers place upon contractors to provide access to project finance as a condition of tender.
3. HERMES, the German export credit institution is not actually government owned but relies upon government support and legislation for most of its operations.
4. In the US case however the majority of tied funds are used for basic aid projects that often do not facilitate the growth of infrastructure and industrial plants. Since

The Financing of International Projects

these types of construction are the specialities of US international contractors, generally American contractors do not benefit as much from tied aid as their French and Japanese counterparts.

5. The Overseas Project Board is an advisory group of industrialists and bankers to the UK government Department of Trade and Industry.

6. The rationale behind import financing was that China was being helped to develop raw materials which Japan would eventually import. In practice the import financing was used to pay for capital equipment from Japan.

7. This is in line with the 'Prisoner's Dilemma' game strategy of oligopoly theory.

Chapter Eight
SUMMARY AND CONCLUSIONS

In Chapter 1 it was stated that the overall objective of the research was to analyse characteristics of international construction that have had major effects on firms' operations within the industry and are also significant to the contemporary market environment. This chapter presents a summary of the findings and conclusions of the analysis presented in the preceding chapters in the context of this objective.

In the next section a brief summary of the major points of the thesis is given, and this is followed by analysis of the major hypotheses of the research which form the basis of the investigation, with reference to the theoretical and empirical findings. The third and fourth sections of the chapter deal with the major conclusions of the research, which have been separated into theoretical and practical issues for clarity though the two are clearly related. Finally limitations of the research and suggestions for future work in this field are given.

8.1 SUMMARY OF THE THESIS

Before outlining the major conclusions of the research it is useful to summarise the major points of the analysis in the form of a chapter-by-chapter synopsis.

Chapter 1 serves as an introduction to the topic and outlines the overall objectives and hypotheses of the research. The chapter additionally puts forward the reasoning behind the use of international production theory as the general framework for the research.

Summary and Conclusions

In Chapter 2 the theoretical background of the thesis is expanded beyond the brief introduction in Chapter 1 to illustrate the major strands of the theory of international production and the multinational enterprise. Given the advantages and disadvantages of the major strands of thought, the chapter points to Dunning's eclectic framework as the most suitable approach to the research because of its generality and ability to combine the various theoretical strands into one integrated paradigm.

In Chapter 3 as an introduction to international construction the special characteristics of the industry are outlined in the context of economic theory, to show how these characteristics have determined the nature of the industry. Additionally those factors that distinguish international construction from domestic construction are discussed to illustrate that international contractors face an environment of greater risk and uncertainty than their domestic counterparts.

Chapter 4 combines the analysis of chapters 2 and 3 to form the major predictions of the application of the eclectic framework to the international construction industry. These predictions serve as the basis for the empirical investigation into the industry outlined in chapters 5 and 6.

On the basis of the questionnaire survey carried out within the research, Chapter 5 illustrates the major O advantages in international construction, which ultimately come down to the exploitation of firm specific and country specific factors. The chapter suggests that contractors from different countries are likely to have different advantages, and that national differences in success overseas may be attributable to the nature of these advantages and the way in which they are exploited.

Chapter 6, as a continuation of the empirical analysis of Chapter 5, analyses internalisation and locational aspects of international construction. The chapter illustrates the usefulness of internalisation theory particularly for policy relating to political risk minimisation and joint venture, but is also useful in explaining the hierarchy of international construction companies. Locational factors discussed tend to emphasise the interaction of O and L advantages in market regions as predicted by the theoretical framework.

In Chapter 7 project finance is introduced as a major source of O advantage, and the concept of financial competition is discussed in detail. The chapter suggests that some nationality contractors have been more successful than others because of the level of financial support they receive from their home government, and that UK contractors may be at a relative disadvantage in this respect because of the limited funds made available to them by the British government.

8.2 RELEVANCE OF THE HYPOTHESES OF THE RESEARCH

In Chapter 1 we outlined three hypotheses which formed the basis of the research as a theoretical and empirical investigation. On the findings and results of this work as summarised above, we now return to the hypotheses to see how accurately they have predicted the major features of the analysis.

The first hypothesis, which states that the construction industry is differentiable from any other because of the unique combination of characteristics in the industry, provides justification for this research is valid. Clearly if the industry does not have any traits which separate it from others, there is no logical reasoning for treating construction as a special case. It was further hypothesised that the problems of the international business environment separate international contracting from domestic contracting so that it may be treated as a separate industry with characteristics which differentiate international construction from domestic construction.

Based on economic analysis, Chapter 3 provides theoretical justification for upholding the first hypothesis. It is argued in this analysis that the nature of the final product in particular has led to the development of the industry as a unique case both in the demand and supply elements of the production process. Additionally it is suggested that the combination of these traits and the problems faced by any enterprise operating across national borders makes the international construction industry differentiable from any other industry, including domestic construction, which is reflected in the competitive environment the contractor faces. According to economic theory the research topic is therefore justified.

The second hypothesis suggests that the theoretical framework of international production and the MNE can be applied to the international construction industry despite the fact that the industry has specific characteristics which

Summary and Conclusions

separate it from others, and also considering that the theory was initially intended as an explanation of multinational manufacturing. For this hypothesis to be validated, clearly it should be possible to integrate the characteristics of the industry with the theoretical framework to provide predictions for empirical testing. Following this reasoning we can see from the theoretical and empirical analysis of chapters 2-7 that this hypothesis may be regarded as justified, and we would additionally argue that the research points to the theory of international production in general and Dunning's eclectic framework in particular as the most flexible (though thorough) method for any industry specific analysis of this type.

As a logical combination of the first two hypotheses, the third hypothesis suggests that if the other hypotheses are justified then the application of the theory to the industry should provide an explanation of the competitive patterns in the industry. In particular firm and country specific explanations of method and motive of overseas operations, competition, and locational policies of the firm. Justification for accepting the third hypothesis stems from the fact that the empirical analysis of the research backs up the theoretical predictions to generate results which are interesting both academically and practically about the state and nature of the industry, as we shall discuss in the following sections concerning theoretical and practical conclusions of the work. The fact that the theoretical framework leads us to these results and conclusions suggests that the overall objective of the research has successfully been achieved.

8.3 THEORETICAL CONCLUSIONS

By theoretical conclusions we refer here to conclusions that we can draw about the adopted theoretical framework from the research. There are four major theoretical conclusions we draw from this work.

The first, and most obvious theoretical conclusion from the discussion above is that the adopted framework is relevant in explaining the behaviour of the international construction industry. Although international production theory in general was initially seen as a theory of multinational manufacturing, the validity of the predictions from the application of Dunning's eclectic theory to the

Summary and Conclusions

industry as outlined in chapters 5 and 6 suggest that the approach may have wider relevance. In the context of the objectives of this research therefore, arguably the most important theoretical conclusion of this research is that the theory of international production can successfully be adopted for industry specific research outside multinational manufacturing, and additionally that the approach may prove useful as a practical guide to future events and policy recommendations in that industry. This is treated in more detail in the following section for the international construction industry.

The second conclusion (which follows on from the first), is that where multinational interests are apparent in an industry, international production theory is probably the most useful tool for determining the so called 'how, where and why' of that industry. However, because of the wide spectrum of interests that are covered by international production theory, general rather than specific approaches to theoretical application should be viewed as the most efficient way of analysing the major features of that multinational industry. The analysis of Chapter 2 and the subsequent empirical research and results of this work testify to the fact that Dunning's eclectic approach is possibly the most flexible and useful tool for this task, regardless of the type of industry analysed. It is hoped that future researchers into multinational industry will bear this in mind when choosing a suitable theoretical framework for industry specific studies.

Although as we have argued the eclectic framework provides a comprehensive and flexible method for analysing the international construction industry despite industry specific characteristics, the application of the framework to the industry illustrates that the eclectic approach is limited to static analysis and would clearly be more acceptable if dynamic considerations were included. This criticism reflects the work of Graham (1986) to a certain degree, who regards the eclectic framework as being unable to cater for oligopolistic strategy of MNEs. By stressing the nature of O advantages Dunning does partially cater for a product cycle type analysis along the lines of Vernon (1966, 1971, 1974) but in this sense the eclectic framework adds little to Vernon's original contribution. However, we would argue that this weakness in Dunning's framework may be disregarded by introducing the Buckley and Casson model of optimal timing of FDI (1981) into the approach, since this

Summary and Conclusions

goes some way to providing explanations of how O advantages may be exploited through time given the nature of internalisation and the market (locational) situation. By stressing the motivations and methods of operating abroad the extended framework suggests why overseas activity is undertaken in the first place and therefore movements from this initial point may be regarded as a change in the relative attractiveness of the components of the framework. Consequently oligopolistic strategy, or other patterns of FDI behaviour, can be investigated within the extended framework through the analysis of OLI components because these elements are likely to be the major impetus behind any changes that do occur. Our third conclusion therefore is that although the Dunning approach may be limited dynamically there is scope for the framework to be extended using the Buckley and Casson model mentioned above.

The fourth theoretical conclusion of this research results from the theoretical background of Chapter 2 and the subsequent empirical analysis, which illustrates that while the industrial organisation and internalisation branches of international production theory have been subject to detailed theoretical analysis, locational factors as an incentive to MNE involvement remain neglected theoretically in this field. We would suggest that as a major determinant of both where and why companies operate overseas, locational aspects of international production have been treated to more empirical investigation rather than theoretical underpinning of why such results should occur. This state is unsatisfactory if we suggest that it is unreasonable to argue that locational factors can be summed up so simply when there is clear interaction between choice of market and what are regarded as the highly complex O and I factors of the eclectic paradigm. The fourth theoretical conclusion of this research is therefore that more theoretical work is needed in the field of locational determinants of MNE activity.

8.4 PRACTICAL CONCLUSIONS

Since the practical conclusions of this research are a product of the interaction of industry specific traits and international production theory, clearly a major feature of the approach is that it is not restricted to academic application. In practical terms for the international contractor the overall conclusion of the research is that it is increasingly the case that the contracting firm must differentiate the services the company offers to win contracts in overseas markets. Because of the need for this type of product differentiation in international construction, and the fact that different locations require different aspects of product differentiation, the international contractor must generate O advantages relevant both to the region the contractor works in and the competitive situation and competitors with which the contractor is faced. Accordingly the research points to five areas of policy that the contractor may incorporate into the firm's strategy and operations to remain competitive:

1. Maintenance of reputation. The construction firm must maintain or enhance its reputation if the company is to be successful in international bidding. The international contracting company should therefore establish the following policies:
   a. Offer construction services that are consistent with the past reputation of the firm so that reputation will correspond with the client's expected level of quality and so maintain or improve the firm's reputation.
   b. Only bid for those projects that can feasibly be carried out to the required standard by the firm.
   c. Maintain an overseas production team which is able to work to the required standard and therefore enhance the reputation of the firm. In practical terms this will involve the establishment of a specific team (or set of teams) that specialises in certain markets and/or technical skills so that they may gain valuable on-the-job experience of conditions and work required.

2. As a general policy, the firm should consciously aim to identify its relative strengths and weaknesses which form the basis of O advantages, and continually update advantages or create new ones. Because of movements in the home and host country economies, client demands, and the international industry through time, it is highly unlikely that O advantages will remain important over time. The contractor should therefore make it an explicit part of the firm's competitive strategy to exploit as many O advantages as possible which can be used to differentiate the firm from others in the industry, and keep these competitive via

Summary and Conclusions

continual market and competitor surveillance.

3. Related to 2, the contractor should aim to compete in the market where full exploitation of the firm's O advantages is likely. It should however be noted that this may not necessarily be the market with the greatest demand. Although the contractor must locate where there is market demand (cf Chapter 3), it may be more profitable to locate in a region where the overall level of demand for construction services is lower than other markets if full exploitation of the firm's O advantages is likely in that market.

4. Because of the nature of the contractor's product and the market conditions contractors are likely to face in developing region markets (ie unpredictable demand conditions and political environment) internalisation theory suggests that it may be beneficial for the contractor to operate in several markets using simultaneous exporting and FDI to diversify risk and also fully exploit the resources of the contractor across markets. Additionally the theoretical framework suggests that where this forms part of a globally coordinated strategy of the firm that contractor will experience transactional advantages on financial and efficiency grounds not available to less coordinated contractors. Global coordination therefore appears the most efficient way of exploiting O advantages within the international construction industry.

5. Contractors from the major developed countries should make more extensive use of joint ventures as a means of obtaining work on those contracts where it would be feasible for several firms to compete efficiently as a single corporation on a contract. This is likely to benefit the contractor by lessening the firm's risk exposure in that country, freeing the company's resources for use on other projects, and possibly creating openings in markets new to the enterprise. However, DC contractors should aim to joint venture with specialised contractors who rely upon a high level of human capital rather than with lesser cost contractors to retain competitive advantage without facing the risk of loss of O advantages.

Although the factors given above stress the firm specific environment, the international contractor must realise that the generation of O advantages is ultimately likely to come down to country specific characteristics. These will not only be reflected in success of the contractor in world markets according to these advantages, but

Summary and Conclusions

additionally may have a positive effect upon transactional advantages. Clearly the more O advantages the contractor can exploit in various markets the greater is the incentive and ability to internalise production worldwide, and therefore the greater are the internal benefits. Ultimately this will lead to global coordination of the company that enables that enterprise to benefit from a coordinated strategy of the firm's resources and objectives worldwide.

Possibly the conclusion of greatest importance to the generation of O advantages which results from this research is that in the contemporary environment in the industry competition has become characterised by the exploitation of country specific factors, so that price may only be considered one feature of competition which is traditionally solely price oriented. The situation has become increasingly dominated by political and historical links, and success in many cases may be put down to the level of political and financial support contractors get from their home government. Furthermore, it is argued here that competition that stresses national characteristics is likely to remain for some time, if not become more prevalent. The success and ability of the French, Korean and increasingly Japanese contractors internationally may be put down to a coordinated national strategy that emphasises national characteristics and political support which has enabled these contractors to remain competitive in the major developing market regions despite aggressive and fierce competition. It is therefore not only likely that these countries' contractors will maintain such a strategy, but that other major contracting countries will follow the example to minimise competition amongst themselves in the international market place. In the context of this argument, it is argued here that there are four major advantages potentially open to UK contractors to help compete abroad and maintain a national entity, although the list is by no means exhaustive.

1. The 'City'. As one of the largest international capital markets in the world, London provides possibly the greatest advantage of all to UK contractors. Clearly in a market dominated by financial competition, the ability and expertise of the City in both capital finance and insurance fields may give UK contractors a competitive advantage where financial competition prevails. However, it should be stressed that the City itself will only be able to obtain capital at market rates of interest and therefore is not likely to provide capital for UK contractors which would be

competitive with highly subsidised official funding of other nationality contractors. The City therefore may provide an extremely useful service, but is unlikely to provide the solution to matching mixed credit financing which is likely to be important in the industry.

2. Related professions. In Chapter 5 we pointed out that many French contractors obtain work through French consultants already advising on projects. Clearly such an arrangement would benefit UK contractors since British consultants are highly demanded in most overseas markets. One objection to this is that UK consultants regard their impartiality as their major competitive advantage and argue that this approach would tarnish that reputation. This may however be something of a false argument since French consultants maintain a high workload regardless of this policy, and would certainly be less tenable the more UK contractors could point to projects carried out to a high standard and quality.

3. Commonwealth links. UK contractors could generate interest in much of the world via ex-empire links. Information about countries, national reputation, UK firms in the market, and close political links with the UK are all exploitable channels that other nationality contractors use to positive effect in their ex-empire markets.

4. UK manufacturing industry. By developing links with UK manufacturers who are involved in international production, the UK construction company can expect to increase its chances of winning bids on overseas facilities provided for those manufacturers. This may give the contractor both an introduction to a new market and enhance reputation both in that market and worldwide.

Although these factors relate to country specific characteristics they are general by nature and may not provide significant differences to the competitive structure of UK contractors without the cooperation of various other institutions, and therefore may not be practically exploitable at all. However from the analysis of chapters 5-7 we argue that there are two particular country specific fields in which UK contractors are lacking and these must change if British contractors are to remain competitive in the industry:

1. Home government support. Although in certain respects the UK government provides extensive services that benefit contractors' overseas operations (eg ECGD services) one area where UK contractors do suffer is the

Summary and Conclusions

level of official finance they may use for mixed credit competition in relation to their major competitors (cf Chapter 7). Although there may be justifiable cases made both for and against the use of mixed credit to obtain work, what is more relevant to this discussion is that competitors of UK contractors are using the mechanism to win contracts in the developing region markets, and therefore to stand any chance of effectively competing in these markets UK contractors must at least be able to match competitors on financial terms. Given the present state of UK government help, this will not only require the government setting aside greater financial resources for mixed credit competition, but also a restructuring of the ATP programme so that UK contractors do not suffer from restrictions in the programme when finance of this type is needed. As an illustration, we recommend that the ATP be revised to allow exporters to use the provision without need for preliminary evidence of competitors' actions, which in many cases makes the programme impractical for the international contractor.

A recommendation to increase the funds to allow more competitive export credit funding is regarded in some parts of the government as the subsidisation of 'lame ducks'. This type of thinking is fundamentally wrong on two points:

a. Similar subsidisation is carried out by all the major contracting countries to generate effective demand rather than create it. The contracting companies that use the funds are competitive and efficient firms which in many cases record profits, and therefore cannot feasibly be classed as 'lame ducks'. Funds are needed because of the nature of project finance rather than the fact that contractors need to create a market for their services.

b. By winning overseas contracts, UK construction firms contribute both directly and indirectly to the UK economy. In a recent Price Waterhouse survey (1985) into the contribution of construction exports to the US economy it was estimated that the indirect impact of construction exports on the economy was a multiplier value of 1.4. (1) If we assume the same multiplier value for the UK (which in practice may have a similar multiplier to the US in this respect) this would mean that in 1984 the £2,381 million of foreign earnings of UK contractors would generate £3,333 million indirectly in the economy. In practice the multiplier is likely to be significantly higher in both economies if direct effects of construction earnings are taken into

Summary and Conclusions

account. If UK contractors are not able to compete because they do not possess the financial resources, this would mean a lower level of revenue to these firms and therefore a smaller positive indirect effect on the economy.

We therefore argue that to maintain a competitive stance in the industry, UK international contractors require greater government support to enable them to compete in those regions where project finance considerations are at a premium.

2. UK contractors. Although the UK government may be criticised for the lack of financial support that has been given to UK contractors in the past, part of the blame for this must be laid upon UK contractors themselves. Unlike many of their competitors UK contractors compete overseas individually rather than as a national group, and so have invariably ended up bidding against each other as well as against other nationality contractors. This has had three negative effects:

    a. Level of government support has had to be limited to the level of help that all UK contractors can receive to prevent the government being accused of bias towards contractors. Because of the limited resources available for financial support this has meant that where two or more UK contractors are competing for the same contract comprehensive financial help cannot be given to any of the firms on equity and efficiency grounds.

    b. UK contractors as a group may have suffered by presenting an individualistic front when competing against effectively nationalised corporations from other countries with full government support.

    c. UK contractors are small compared to many of their foreign competitors and therefore lose out by not having the financial resources to bid on all major projects or sustain losses in risky markets. This leaves these contractors at a significant disadvantage to competitors, although this problem could be alleviated through joint venture or consortium operations that could also reduce the problems outlined in a and b above.

We therefore argue that if UK contractors are to realistically compete overseas with full government support they must adopt a strategy similar to their competitors by competing on a national basis to prevent wasteful competition amongst themselves.

The conclusion that can be drawn from these observations of the failings of the UK construction industry is that the international construction industry is no longer characterised by the competitive firm, but that competition is increasingly becoming reliant upon nationalistic non-market factors, especially home government support. We have hypothesised here that this situation is likely to become more prevalent in the industry as the national entity becomes more common, and therefore to remain competitive UK contractors must follow this trend by also looking at various overseas markets as part of the overall strategy of global competition, that requires global coordination. The consequences of UK contractors and the government not adopting this type of competitive thinking may not only be a fall in overseas earnings, but ultimately may lead to a serious threat to the domestic construction industry in the UK. Japanese contractors for example are currently beginning to compete for work against established indigenous contractors in both Australia and USA on the basis of the experience that they have acquired in the international market. In 1984 for example the US was the largest national overseas market for Japanese contractors surpassing all of the more traditional markets of South East Asia - Australia was the fourth largest market for Japanese contractors in the same year. It is no secret that the European market is also targeted, and it would be both naive and myopic to suggest that Japanese contractors cannot do in the UK what Japanese manufacturers have already done. UK contractors, with the help of the government, must therefore consider both the international and domestic markets as part of their global strategy, and accordingly must remain competitive overseas to maintain domestic interests.

8.5 LIMITATIONS OF THE RESEARCH AND SUGGESTIONS FOR FUTURE WORK

In Chapter 1 we argued that although literature can be found on the workings and influences of international contractors in overseas markets, very little work has been carried out into the 'why, how and where' of multinational involvement of international contractors. In an attempt to partially rectify this, the aim of this research has been to address these factors through economic theory.

Summary and Conclusions

Although the research has in the main achieved all the aims of this investigation, the nature of the work and the subject matter has meant that the analysis had had to be of a general nature so that the major elements of the competitive situation could be outlined. This generality has meant that several interesting topics have not received the detailed analysis they deserve, which we suggest is possibly the greatest limitation of this work. We therefore highlight some of the major issues here as a guide to future research in this field.

1. Joint venture. Although joint venture has been discussed in some depth here, more work is needed to illustrate both the advantages and disadvantages of joint venturing in the pre-bid, bid, and construction phases of the production process especially with respect to joint venture with local and foreign partners. In addition to this, similar work is required for the role of consortia in international construction, particularly with reference to the benefits and costs of such a venture and the criteria behind the emergence and successful operations of consortia.

2. Sub contracting. This study has concentrated mainly upon major project contractors and therefore tended to ignore sub contracting in international construction. However, the subject deserves greater attention because of the high incidence of sub contracting in the industry, so that future work should aim to incorporate analysis on the determinants of sub contracting in the industry and how this affects the operations and returns of the major contractor.

3. Technology transfer. Given that the majority of international contractors work in the developing region markets, technology transfer in the industry needs a great deal more analysis from the point of view of both the contracting firm and the host country government and clients. It is suggested that work in this field is required to remove or at least lessen some of the controversy that surrounds the subject in the industry, so that technology transfer policies can be formulated which are agreeable to both the contractor and the host country government.

4. Macro economic variables. Although in the research we have outlined the differences between the major countries' contractors in detail in terms of level of government support, it has been beyond the scope of this work to analyse the importance of macro economic variables such as foreign exchange movements and interest rate differentials on the competitiveness of contractors

overseas. In that this may provide a competitive edge to one nationality contractor over contractors from other countries, the subject should be treated to economic, and more probably econometric, analysis.

5. Construction related services. The influence of service enterprises that influence contractors' overseas operations has been found in this work to be a significant feature of the competitiveness of certain contractors' overseas operations, particularly in the construction consulting, financial, and (in certain cases) construction materials sectors. More work is required upon the actual effect of these factors on the competitiveness of contractors abroad, and related to this more detailed analysis is required to discern which of these factors contractors have internalised as part of a vertically diversified strategy. Ultimately this is likely to reflect the relevance of the globally integrated strategy to international contractors.

The overall recommendation for future work in this field is that the whole of this research topic needs extending past that presented in this thesis. We have pointed here to some of the major differences between competitors from the major contracting countries, and suggest that this work needs greater empirical investigation of these contractors' operations and competitive advantages. In addition to this similar research is needed into international contracting firms from LDCs. We argue here that this work is necessary for contractors and academics alike to get a clearer picture of the industry so that those aspects that require greater research (of which the list above is only a small part) will not be restricted by lack of knowledge about the major competitors in the industry, and the environment in which they compete. The industry has for the most part been ignored, despite the fact that it generates billions of dollars of revenue each year, is a major source of employment directly and indirectly in both the developing and developed countries, and perhaps most importantly provides the infrastructure for development through social and industrial improvement in many areas of the world. It should therefore be the overall objective of future work in this field to amend this shortcoming.

Summary and Conclusions

8.6 FOOTNOTE

1. That is, $1 of direct revenue in construction generates $1.40 of indirect revenue in other sectors.

REFERENCES

1. Aharoni, Y. (1966) The Foreign Investment Decision Process (Boston, Massachusetts, Harvard University Press)
2. Alchian, A. and H. Demsetz (1972) 'Production, Information Costs and Economic Organisation' American Economic Review, 62 (December)
3. Aliber, R.Z. (1970) 'A Theory of Direct Foreign Investment' in C.P. Kindleberger (ed.) The International Corporation: A Symposium (Cambridge, Massachusetts: MIT Press), chapter 1
4. Aliber, R.Z. (1971) 'The Multinational Enterprise in a Multiple Currency World' in J.H. Dunning (ed.) The Multinational Enterprise (London: Allen and Unwin)
5. ANCE (1985) L'industria delle costruzioni nel 1984 (Rome: ANCE)
6. Ashworth, A. (1983) 'Contractual Methods used in the Construction Industry' Building Trades Journal, May 12th
7. Association of Consulting Engineers (ACE) (1985) Annual Report to Members
8. Atsumi, T. (1982) 'Present Status of Japanese Construction Industry' Constructor, January
9. Bain, J.S. (1956) Barriers to New Competition (Boston, Massachusetts: Harvard University Press)
10. Baker, M.R. and R.W. Cockfield (1984) 'International Turnkey Projects - The Risks and Methods of Impact Reduction' Fourth International Symposium on Organisation and Management of Construction, Chartered Institute of Building pp.17-29
11. Barna, T. (1983) 'Process Plant Contracting: A Competitive New European Industry' in G. Shepherd, F.

References

Duchene and C. Saunders (eds.) European Industries Public and Private Strategies for Change (London: Frances Pinter)

12. Barna, T., J. Aylen, G. Dosi and D. Jones (1981) The European Process Plant Industry, unpublished report to the Commission of the European Communities (Brighton: Sussex European Research Centre)

13. Buckley, P.J. (1983) 'New Theories of International Business: Some Unresolved Issues' in M.C. Casson (ed.) The Growth of International Business (London: Allen and Unwin)

14. Buckley, P.J. (1985) 'A Critical View of Theories of the Multinational Enterprise' in P.J. Buckley and M.C. Casson (eds.) The Economic Theory of the Multinational Enterprise (London: Macmillan)

15. Buckley, P.J. and M.C. Casson (1976) The Future of the Multinational Enterprise (London: Macmillan)

16. Buckley, P.J. and M.C. Casson (1981) 'The Optimal Timing of a Foreign Direct Investment' Economic Journal, 91 (March)

17. Buckley, P.J. and H. Davies (1979) 'The Place of Licensing in the Theory and Practice of Foreign Operations' University of Reading Discussion Papers in International Investment and Business Studies, 47

18. Buckley, P.J. and P. Enderwick (1985) 'Manpower Management in the Domestic and International Construction Industry', mimeo, Bradford: University of Bradford Management Centre, and Belfast: University of Belfast Department of Economics

19. Calvet, A.L. (1981) 'A Synthesis of Foreign Direct Investment Theories and Theories of the Multinational Firm' Journal of International Business Studies, 12 pp.43-60

20. Casson, M.C. (1979) Alternatives to the Multinational Enterprise (London: Macmillan)

21. Casson, M.C. (1982) 'Transactions Costs and the Theory of the Multinational Enterprise' in A.M. Rugman (ed.) New Perspectives in International Business (London: Croom Helm)

22. Casson, M.C. (1983) 'Introduction: the Conceptual Framework' in M.C. Casson (ed.) The Growth of International Business (London: Allen and Unwin) chapter 1.

23. Casson, M.C. (1984) 'General Theories of the MNE: A Critical Examination' University of Reading Discussion Papers in International Investment and Business Studies, 77

24. Casson, M.C. (1985a) 'Foreign Investment and

References

Economic Warfare: Internalising the Implementation of Threats', mimeo, University of Reading

25. Casson, M.C. (1985b) 'Alternative Contractual Arrangements for Technology Transfer: New Evidence from Business History', mimeo, reproduced as University of Reading Discussion Papers in International Investment and Business Studies, 95 (May 1986)

26. Casson, M.C. (1985c) 'Diversification and Subcontracting' forthcoming in P.M. Hillebrandt, W.D. Biggs and H.J. Dunning Management Strategy of Large UK Contracting Firms: theory and practice (London: Macmillan, 1988)

27. Casson, M.C. (1986) 'Vertical Integration and Intra-Firm Trade' in M.C. Casson (ed.) Multinationals and World trade (London: Allen and Unwin)

28. Casson, M.C. and G. Norman (1983) 'Pricing and Sourcing Strategies in a Multinational Oligopoly' in M.C. Casson (ed.) The Growth of International Business (London: Allen and Unwin)

29. Caves, R.E. (1971) 'International Corporations: the Industrial Economics of Foreign Investment' Economica, 38 (February) pp. 1-27

30. Caves, R.E. (1974) 'Industrial Organisation' in J.H. Dunning (ed.) Economic Analysis and the Multinational Enterprise (London: Allen and Unwin), chapter 5

31. Caves, R.E. (1982) Multinational Enterprise and Economic Analysis (Cambridge: Cambridge University Press)

32. Central Statistical Office (1984) United Kingdom Balance of Payments 1984 Edition (London: HMSO)

33. Cezik, O.A. (1980) 'Across the Bridge - New markets?' Building Technology and Management, May pp.13-14

34. Cheney, D.M. (1985) 'The OECD Export Credits Agreement' Finance and Development, September

35. CIM (1985) Costruttori Italiani nel Mondo (Rome: ANCE)

36. Civil Engineering (1979) 'Italy Booms Abroad', November p.19

37. Coase, R. (1937) 'The Nature of the Firm' Economica, 4 (November)

38. Cockburn, C. (1970) Construction in Overseas Development Overseas Development Institute Ltd., London

39. Construction Association of Korea (1985) The Construction Association of Korea (Seoul: CAK)

40. Cox, V.L. (1982) International Construction

References

(London: Construction Press)

41. Demacopoulos, A. and F. Moavenzadeh (1985) International Construction Financing TDP Report 85-3 (Cambridge, Massachusetts: MIT)
42. Department of the Environment (1985) Housing and Construction Statistics (London: HMSO)
43. Dun and Bradstreet (1985) Who Owns Whom 1984 (London: Unwin Brothers Ltd)
44. Dunn, A. and M. Knight (1982) Export Finance (London: Euromoney)
45. Dunning, J.H. (ed.) (1974) Economic Analysis and the Multinational Enterprise (London: Allen and Unwin), chapter 1
46. Dunning, J.H. (1977) 'Trade, Location of Economic Activity and the Multinational Enterprise: A Search for an Eclectic Approach' in B. Ohlin, P.O. Hesselborn and P.M. Wijkman (eds.) The International Allocation of Economic Activity (London: Macmillan) pp.395-418
47. Dunning, J.H. (1979) 'Explaining Changing Patterns of International Production: In defence of the Eclectic Theory' Oxford Bulletin of Economics and Statistics, 41 pp.269-95
48. Dunning, J.H. (1980) 'Towards an Eclectic Theory of International Production' Journal of International Business Studies, 11 pp.9-31
49. Dunning, J.H. (1981) International Production and the Multinational Enterprise (London: Allen and Unwin)
50. Dunning, J.H. (1982) 'Non-equity Forms of Foreign Economic Involvement and the Theory of International Production' University of Reading Discussion Papers in International Investment and Business Studies, 59
51. Dunning, J.H. (1985) 'The Eclectic Paradigm of International Production: an update and Reply to its Critics' op.cit., 90
52. Dunning, J.H. and J.A. Cantwell (1982) 'Joint Ventures and Non-equity Foreign Involvement by British Firms with particular reference to Developing Countries: an Exploratory Study' op.cit., 68
53. Dunning, J.H. and M. McQueen (1981) 'The Eclectic Theory of International Production: a case study of the International Hotel Industry' Managerial and Decision Economics, 2 pp.197-210
54. Dunning, J.H. and G. Norman (1979) Factors influencing the Location of Offices of Multinational Enterprises in the UK (London: Economists Advisory Group)

References

55. Dunning, J.H. and R.D. Pearce (1981) The World's Largest Industrial Enterprises (Farnborough: Gower)
56. Edmonds, G.A. (1972) 'Role of Construction Industry in Overseas Development' Civil Engineering and Public Works Review, June pp.624-625
57. Eiteman, D.K. and A.I. Stonehill (1982) Multinational Business Finance (Reading, Massachusetts: Addison-Wesley)
58. Engineering News-Record (1979) 'US Arrogance costs firms billions in lost jobs' 29th November pp.26-37
59. Engineering News-Record (1981) 'Joint ventures win big contracts' 30th April, pp.25-26
60. Engineering News-Record (July 1982) 'Foreign markets grow despite recession' 15th July, pp.54-79
61. Engineering News-Record (Sept. 1982) 'Banks still lending - cautiously' 16th September, pp.10-11
62. Engineering News-Record (1983) 'Recession cuts foreign work almost 9%' 21st July, pp.50-74
63. Engineering News-Record (Aug. 1984) 'Foreign billings up 3% for 1983' 2nd August, pp.36-47
64. Engineering News-Record (Sept. 1984) 'Meticulous plans make plants sail' 27th Sept. pp.24-29
65. Engineering News-Record (Aug. 1984) 'Financial engineering wins jobs' 2nd August, pp.30-35
66. Engineering News-Record (1985) 'Foreign contracts slump further' 18th July pp.34-58
67. Euromoney (1982) 'Who does the work in a loan syndication', March (London: Euromoney)
68. European Community Contractors (1984) Construction in Europe Statistics 1970-83, (Paris: EEC)
69. Export Group for the Construction Industries (1979) Major British Construction Contractors
70. Federation Nationale du Batiment (1983) Travaux de Batiment a l'etranger en 1983
71. Financial Times (Jan. 1980) 'Arab Construction Survey' 22nd January, pp.I-X
72. Financial Times (Sept. 1980) 'International Construction' 24th September, pp.I-VI
73. Financial Times (1981) 'International Construction' 24th August pp.I-VI
74. Financial Times (1984) 'International Construction' 5th March, pp.13-18
75. Financial Times (1985) 'Slowdown in the Middle East Construction Upsurge' 20th May p.17
76. Gabriel, P.P. (1967) The International Transfer of

References

Corporate Skills: Management Contracts in Less Developed Countries (Boston: Harvard University)

77. Graham, E.M. (1974) Oligopolistic Imitation and European Direct Investment in the United States, PhD thesis Harvard University

78. Graham, E.M. (1978) 'Transatlantic Investment by Multinational Firms: a Rivalistic Phenomenon?' Journal of Post Keynesian Economics, 1 pp.82-99

79. Graham, E.M. (1985) 'Intra industry Direct Foreign Investment, Market Structure, Firm Rivalry and Technology Performance' in A. Erdilek (ed.) Multinationals as Mutual Invaders: Intra industry Direct Foreign Investment (London: Croom Helm)

80. Graham, E.M. (1986) 'Internal Economies, Oligopoly Reaction, and Dynamic Contestability in Global Industries: A First Cut at Synthesis', mimeo, University of North Carolina

81. Hamman, N.N. (1971) Contracting Firms and the Pre Tendering Phase of International Construction Projects, unpublished MSc thesis, UMIST

82. Harrison, R.S. (1982) 'Tendering Policy and Strategy' in R.A. Burgess (ed.) Construction Projects: their Financial Policy and Control (London: Construction Press)

83. Hauptverband der Deutschen Bauindustrie (1985) 'Baukonjuntur-Spiegel' 10th May

84. Hawker Siddeley Group (1985) Private Analysis summarised in Financial Times 'How Britain is 'being gazumped' in world markets' April 6th

85. Hennart, J-F (1982) A Theory of Multinational Enterprise (Ann Arbor: University of Michigan Press)

86. Hillebrandt, P.M. (1985) Economic Theory and the Construction Industry (London: Macmillan)

87. Hirsch, S. (1976) 'An International Trade and Investment Theory of the Firm' Oxford Economic Papers, 28 (July) pp.258-69

88. Hymer, S.H. (1976) The International Operations of National Firms: A study of Direct Foreign Investment PhD thesis (1960), published 1976 (Cambridge Massachusetts: MIT Press)

89. International Monetary Fund (1985) International Financial Statistics (Washington: IMF)

90. Japanese Ministry of Construction (1985) Japanese Overseas Construction July

91. Johanson, J. and J-E. Vahlne (1977) 'Internationalisation of the firm' Journal of International

References

Business Studies, Spring/Summer

92. Johnson, H.G. (1970) 'The Efficiency and Welfare Implications of the Multinational Corporation' in C.P. Kindleberger (ed.) The International Corporation: a Symposium (Cambridge Massachusetts: MIT Press), chapter 2

93. Kaldor, N. (1934) 'The Equilibrium of the Firm' Economic Journal, 4 pp.60-76

94. Kemp, M.C. (1962) 'The Benefits and Costs of Private Investment from Abroad: a Comment' Economic Record, 38 pp.108-10

95. Kim, S.H. (1983) International Business Finance (Virginia: Robert F. Dame, Inc.)

96. Kindleberger, C.P. (1969) American Business Abroad: Six Lectures on Direct Investment (New Haven: Yale University Press)

97. Knickerbocker, F.T. (1973) Oligopolistic Reaction and Multinational Enterprise (Boston, Massachusetts: Harvard University Press)

98. Kogut, B. (1985) 'Designing Global Strategies: Profiting from Operational Flexibility' Sloan Management Review, Fall, Vol. 27 No. 1

99. Kojima, K. (1978) Direct Foreign Investment (London: Croom Helm)

100. Kompass (1984) Company Information United Kingdom 1984 22nd Edition (East Grinstead: Kompass Publishers Ltd.)

101. Lall, S. (1980) 'Monopolistic Advantages and Foreign Involvement by US Manufacturing Industry' Oxford Economic Papers, 32 (March) pp.102-22

102. Lancaster, K. (1966) 'A New Approach to Consumer Theory' Journal of Political Economy, April pp.132-57

103. Lancaster, K. (1966a) 'Change and Innovation in the Technology of Consumption' American Economic Review Papers and Proceedings, May pp.14-23

104. Lessard, D.R. (1976) 'The Structure of Returns and Gains from International Diversification' in N. Elton and M. Gruber International Capital Markets (Amsterdam: North-Holland)

105. Lessard, D.R. (1977) 'International Diversification and Direct Foreign Investment' in D.K. Eiteman and A.I. Stonehill Multinational Business Finance (published 1982) (Reading Massachusetts: Addison-Wesley)

106. Lessard, D.R. (1979) 'Transfer Prices, Taxes and

References

Financial Markets: Implications of Internal Financial Transfers within the Multinational Firm' in R.G. Hawkins (ed.) Economic Issues of Multinational Firms (Greenwich, Connecticut: JAI Press)

107. Luostarinen, R. (1978) 'The Impact of Physical, Cultural and Economic Distance on the Geographical Structure of the Internationalization Pattern of the Firm' Helsinki School of Economics, FIBO Working Paper No. 1978/2

108. MacDougall, G.D.A. (1960) 'The Benefits and Costs of Private Investment from Abroad: A Theoretical Approach' Economic Record, 36 pp.13-35

109. Magee, S.P. (1976) 'Technology and the Appropriability Theory of the Multinational Corporation', mimeo, reproduced in J. Bhagwati (ed.) The New International Economic Order: the North-South Debate (1977) (Cambridge, Massachusetts: MIT Press)

110. Magee, S.P. (1977) 'Multinational Corporations, the Industry Technology Cycle and Development' Journal of World Trade Law, 11 (July/August) pp.297-321

111. Mason, R.H., R.R. Miller and D.R. Weigel (1975) International Business (New York: John Wiley and Sons Inc.)

112. Markowitz, H. (1952) 'The Utility of Wealth' Journal of Political Economy, 60 pp.151-158

113. MEED (1985) Middle East Directory Contracts Analysis 1984 First Half (London: MEED Consultants)

114. NEDO (1978) Design and Export, Civil Engineering EDC Report (London: HMSO)

115. Neo, R.B. (1975) 'International Construction Contracting' (Farnborough: Gower)

116. Nicholson (1982) Construction and the Common Market, Report to EEC Commission (Paris: EEC)

117. Norman, G. and J.H. Dunning (1983) 'Intra industry Production as a form of International Economic Involvement: an Explanatory Paper' University of Reading Discussion Papers in International Investment and Business Studies, 74

118. OECD (1976) The Export Credit Financing Systems in OECD Member Countries (Paris: OECD)

119. OECD (1984) Development Cooperation 1984 Review (Paris: OECD)

120. Overseas Project Board (1985) Fourth Report, British Overseas Trade Board (London: HMSO)

121. Ozawa, T. (1979) 'International Investment and Industrial Structure: New Theoretical Implications from the

References

Japanese Experience' Oxford Economic Papers, 31 (March)

122. Park, W.R. (1970) The Strategy of Contracting for Profit (New Jersey: Prentice-Hall Inc.)

123. Pearce, J. (1980) 'Subsidised Export Credit' Chatham House Papers, No. 9 (London: The Royal Institute of International Affairs)

124. Penrose, E.T. (1956) 'Foreign Investment and the Growth of the Firm' Economic Journal, 66 (June) pp.220-35

125. Porter, M.E. (1980) Competitive Strategy - Techniques for Analysing Industries and Competitors (New York: Free Press)

126. Price Waterhouse (1985) The Contribution of Architectural, Engineering and Construction Exports to the US Economy (Washington: Price Waterhouse)

127. Robbins, S.M. and R.B. Stobaugh (1974) Money in the Multinational Enterprise (New York: Basic Books)

128. Rugman, A.M. (1976) 'Risk Reduction by International Diversification' Journal of International Business Studies, 7 pp.75-80

129. Rugman, A.M. (1977) 'Risk, Direct Investment and International Diversification' Weltwirtschaftliches Archiv, 113 pp.487-500

130. Rugman, A.M. (1979) International Diversification and the Multinational Enterprise (Farnborough: Lexington)

131. Rugman, A.M. (1980) 'Internalisation as a General Theory of Foreign Direct Investment' Weltwirtschaftliches Archiv, 116 (June)

132. Rugman, A.M. (1981) Inside the Multinationals (London: Croom Helm)

133. Stevens, G.V. (1974) 'Determinants of Investment' in J.H. Dunning, op.cit., chapter 3

134. Stopford, J.M. and L.T. Wells, Jr (1972) Managing the Multinational Enterprise: Organisation of the Firm and Ownership of the Subsidiaries (New York: Basic Books)

135. Sung, Hwan-Jo (1982) 'Overseas Direct Investment by South Korean Firms: Direction and Pattern' in K. Kumar and M.G. McLeod (eds.) Multinationals from Developing Countries (Massachusetts: Lexington Books)

136. Swedenborg, B. (1979) The Multinational Operations of Swedish Firms: an Analysis of Determinants and Effects (Stockholm: Almqvist and Wiksell International)

137. Teece, D.J. (1976) The Multinational Corporation and the Resource Cost of International Technology Transfer (Cambridge, Massachusetts: Ballinger)

138. Teece, D.J. (1981) 'The Multinational Enterprise:

References

Market Failure and Market Power Considerations' Sloan Management Review, Vol. 22 No. 3 (Spring) pp.3-17

139. Teece, D.J. (1982) 'A Transactions Cost Theory of the Multinational Enterprise' University of Reading Discussion Papers in International Investment and Business Studies, 66

140. Teece, D.J. (1983) 'Technology and Organisational Factors in the Theory of the Multinational Enterprise' in M.C. Casson (ed.) The Growth of International Business (London: Allen and Unwin)

141. The Times (1985) The Times 1000 1984-85 (London: Times Books)

142. Turkish Daily News (1985) Contracting Supplement, 16th June

143. TUSIAD (1984) The Turkish Economy 1984 (Istanbul: TUSIAD)

144. Umeda, K. (1980) The Changing Structure of the Japanese Construction Industry, report to Euro-Construct (European Construction Group)

145. Vernon, R. (1966) 'International Investment and International Trade in the Product Cycle' Quarterly Journal of Economics, 80 (May) pp.190-207

146. Vernon, R. (1971) Sovereignty at Bay: the Multinational Spread of US Enterprises (New York: Basic Books)

147. Vernon, R. (1974) 'The Location of Industry' in J.H. Dunning, op.cit., chapter 4

148. Vernon, R. (1983) Organisational and Institutional Responses to International Risk, Mimeo

149. Westphal, L.E., Y.W. Rhee, L. Kim and A. Amsden (1984) Exports of Capital Goods and Related Services from the Republic of Korea, World Bank Staff Working Papers, No. 629 (Washington: The World Bank)

150. Williamson, O.E. (1971) 'The Vertical Integration of Production Market Failure Considerations' American Economic Review, 61 (May)

151. Williamson, O.E. (1975) Markets and Hierarchies: Analysis and Antitrust Implications (New York: Free Press)

152. Williamson, O.E. (1979) 'Transaction Cost Economics: the Governance of Contractual Relations' Journal of Law and Economics, 22 (October)

153. World Bank (1984) World Development Report 1984 (Oxford: Oxford University Press)

154. World Construction (1979) 'For Korean Contractors - A year of Change', June

References

155. Yannopoulos, G.N. (1983) 'The Growth of Transnational Banking' in M.C. Casson (ed.) The Growth of International Business (London: Allen and Unwin)

BIBLIOGRAPHY

1. Abbott, P. (1981) 'The Next Twenty Years' Civil Engineering, July
2. Ashley, S. (1982) 'Hong Kong Services - How they Build' Building Services, August
3. Baum, W.C. (1970) 'The Project Cycle' Finance and Development, June
4. Brander, J.A. (1981) 'Intra-Industry Trade in Identical Commodities' Journal of International Economics, 11
5. British Business (1983) 'Building up', 28th October
6. British Business (March 1984) 'Construction Output Up', 23rd March
7. British Business (Aug. 1984) 'Designs Abroad: UK Consultants Work Overseas', 24th August
8. British Business (October 1984) 'Building Success Overseas: UK Contractors continue to win Major Orders', 26th October
9. Building (1984) 'Exports to China', 29th June
10. Chadenet, B. and J.A. King, Jr. (1972) 'What is a World Bank Project?' Finance and Development, September
11. Civil Engineering (1978) 'Sources of Finance for the International Contractor', October
12. Civil Engineering (1983) 'Is the UK Market Share set to Fall?' October
13. Colaco, F.X. (1985) 'International Capital Flows and Economic Development' Finance and Development, September
14. Contract Journal (April 1983) 'UK comes last in International Construction Investment League', 21st April

15. Contract Journal (Nov. 1983) 'Africa eclipses Middle East as top Overseas Market' 17th November
16. Corden, W.M. (1974) 'The Theory of International Trade' in J.H. Dunning, op.cit., chapter 7
17. Crosthwaite, C.D. (1972) 'Arrangement of Finance for Overseas Construction Projects' Civil Engineering and Public Works Review, April
18. Davies, M.P. (1972) 'Finance for Civil Engineering Projects' op.cit.
19. Eminton, S. (1985) 'British Know-How helps improve Cairo's Sewerage System' Water Bulletin, 8th November
20. Financial Times (1985) 'British wrangle over Aid for Trade' 24th May
21. Flanagan, R., C. Gray, G. Norman and H. Seymour (1985) 'Strangers in a Strange Land' Building, 15th March
22. Gaylor, R.F. (1978) 'Building in Saudi Arabia' Building Technology and Management, June
23. Gruber, W., D. Mehta and R. Vernon (1967) 'The R&D Factor in International Trade and International Investment in the United States' Journal of Political Economy, 75 (February)
24. Halsey, D. (1979) 'Opportunities Abound in the Far East' Civil Engineering and Public Works Review, March
25. Hertz, D.B. (1965) 'Risk Analysis in Capital Investment' Harvard Business Review, 54 (April)
26. Horst, T.O. (1974) 'The Theory of the Firm' in J.H. Dunning, op.cit., chapter 2
27. Krugman, P. (1979) 'Increasing Returns, Monopolistic Competition and International Trade' Journal of International Economics, 9
28. Lee, Jae-joon (1982) 'The Construction Association of Korea' Constructor, January
29. Lunn, P. (1973) 'Medium and Long Term Export Credit' Journal of the Institute of Bankers, 94 (December)
30. McQuade, L.C. (1975) 'Petrochemical Plant Construction and the Oil Exporting States' Colombia Journal of World Business, Summer
31. Norman, G. and J.H. Dunning (1983) 'Intra-Industry Foreign Direct Investment: its Rationale and Trade Effects', mimeo, University of Reading
32. Norman, G. and N. Nichols (1982) 'Dynamic Market Strategy under threat of Competitive Entry: an Analysis of the Pricing and Production Policies open to the

Bibliography

Multinational Company' The Journal of Industrial Economics, September/December

33. Olsen, B.O.M. (1979) 'Building and Exports' Building Research and Practice, July/August

34. Oman, C. (1984) New Forms of International Investment in Developing Countries (Paris: OECD)

35. Parry, T.G. (1973) 'The International Firm and National Economic Policy: a Survey of some Issues' Economic Journal, 83 (December)

36. Phang Yee Kee (1984) A Study of the growth of South Korea's Overseas Construction, and in Relation to the Behaviour of International Construction Firms, to assess the Problems and Prospects of the Construction Industry in Malaysia, unpublished MSc thesis University College London

37. Ross, I.L. (1972) 'Overseas Contracting and Finance' Civil Engineering and Public Works Review, June

38. Seymour, H. (1985) 'The International Construction Industry 1980-83' Proceedings of the 7th Bartlett International Summer School, Vaulx-en-Velin 1985 (London: Bartlett International Summer School, 1986)

39. Seymour, H., R. Flanagan, and G. Norman (1985) 'International Investment in the Construction Industry: an Application of the Eclectic Approach' University of Reading Discussion Papers in International Investment and Business Studies, 87

40. Seymour, H., J.H. Dunning, and G. Norman (1986) 'Management Strategy in the International Construction Industry' forthcoming in P.M. Hillebrandt, W.D. Biggs and J.H. Dunning, op.cit.

41. Thompson, P. (1980) Construction in the 1980s, unpublished MBA dissertation, University of Bradford

42. Westphal, L.E. (1978) 'The Republic of Korea's experience with Export-led Industrial Development' World Development, Vol. 6, No. 3

43. Williams and Glyn's Bank (1981) 'Export Finance' Quantity Surveyor, May

INDEX

Africa 218-21
  and French contractors 6, 219
aid and trade 250

bilateral aid 252-5
  see also export credit
bonding arrangements 141-2

'Chaebol' 155
comparative advantage 51-4
  see also ownership advantages
construction industry
  barriers to entry 63, 75-6
  capital requirements 63
  demand for the firm 71-2
    and location 72
  demand for services 59-61, 68
  final product 69-70
  industrial structure 74-8
  risk exposure 71
  specialist expertise 72
  supply determinants 73-4
countertrade 168
country specific ownership advantages 143-76
  size of domestic market 146-9

comparative advantage 149-58
consultancy links 158-61
home government support 161-76

design and build 77

eclectic paradigm 264
  see also Dunning, J.H.

export credit
  as a competitive advantage 240-1
  insurance 238-9
  financing 239-40
  mixed credit financing 248-56
    see also bilateral aid

firm specific ownership advantage 139-43
  quality of human capital 140-1
  reputation and name 139, 266
  size of firm 141-3
foreign direct investment (FD)
  definition 16

Index

differences from portfolio inv. 17
mobility of competitive advantages 22
other forms of foreign involvement 47-8
the product cycle
French contractors 141, 151, 153, 171, 268
export credit 242-4
and French consultants 158
and government support 164, 172
tied aid 242-4
see also Africa

German contractors 145, 154, 172
export credit 242-4
and government support 164, 172
global strategy 226-7

Hecksher - Ohlin model 17
horizontal integration 124-6

internalisation theory 55-7
appropriability theory 39-40
contractual benefits and costs 34-6
exporting 46
licensing 46
in international construction 178
and location factors 37
and the multinational enterprise 33-9
oligopolistic strategy 41-3
risk diversification 40-1
transactions costs 43-6
international construction
barriers to entry 99-100
definition of term 9

differences from national contracting 12, 78-84
internalisation factors 106-26, 179-89
exporting/FDI 112, 183
modal choice 107-9
non-equity involvement 113-16
subsidiary operations 182
locational factors 100-5
home and host government links 105
market demand 104
oligopolistic reaction 99-100
product differentiation 89-92
additional incentives 91
by product 90
motives 89
through bidding 89
through firm's name and reputation 89
intra industry foreign direct investment (IIFDI) 30
Italian contractors 145, 154, 172-3
export credit 242-4
and government support 164, 173

Japanese contractors 145, 153, 155, 173, 268, 272
export credit 242-4
and government support 165, 174
see also Soga-Shosha

Kojima theory 51-4, 223-5
and international construction 223-5
and Japanese contractors 224
and Korean contractors 223-4

Index

Korean overseas construction corporation (KOCC) 168

Latin America 221-3
location factors 55-7, 265
  comparative advantage 101
  country specific factors 210-25
  and internationalisation theory 37, 38
  market size 49
  of the individual contractor 197-210
    choice of market 198-205
    incentives/disincentives 203-4
  structural market imperfections 100
  tariff walls 50
  transfer pricing 50
management contracting 47-8
Middle East 210-15
multinational enterprise (MNE)
  barriers to entry 20
  currency differentials 23-5
  financing operations
    internal sources 233-4
    external sources 234-7
    general theories 54-7
  in alien markets 20
  oligopolistic strategy 28-31
  product differentiation 25-6
  and internalisation theory 31-9
  and location theory 48-53
  and market imperfections 21

non-equity involvement 113-20
  and transactions costs 119-20

oligopolistic reaction
  in international construction 99-100, 229-30
ownership advantages 55-6, 85, 87
  and structural market imperfections 87
  and transactionsal market imperfections 87-8
  see also country specific ownership advantages and firm specific ownership advantages

political risk 82
  and global strategy 226
  joint ventures 225
  risk minimisation 225
post supplied construction 59-60
pre-demanded construction 59-61
process plant construction 9
product differentiation 129-32, 266
  against international contractors 133-4
  against local contractors 131-3
  by services offered 134-40
project finance 235-7
psychic distance 44

quality control 44, 116-17

raw materials 61-2

selection of contractor 65-8
  negotiation 66

Index

tendering
  open 66
  selective 66
  two-stage 66
shareholders' interests 198
Société Navale Française de Formation et de Conseil 161
'Soga Shosha' 155
  <u>see also</u> Japanese contractors
South East Asia 215-18
  China 215
South Korean contractors 2, 145, 149, 153, 155, 174, 268
  export credit 242-4
  and government support 165, 174
  and joint venture 193
  in the Middle East 214
sub-contracting 64

tied aid <u>see</u> export credit
transfer pricing 50
Turkish contractors 145, 175
  and government support 175
  in the Middle East 214
turnkey deals 77

UK contractors 145, 170, 268-72
  export credit 242-4, 270
  and government support 165, 170, 269
  invisible earnings 3
  and UK consultants 158, 269-71
US Army Corps of Engineers (USACE) 160-1
US contractors 145, 153, 170
  export credit 242-4
  and government support 165, 170
  and US consultants 158

vertical integration 122-4, 195-6

AUTHOR INDEX

Ahroni, Y. 19
Alchian, A. & H. Demsetz 32
Aliber, R.Z. 21, 23-5, 87, 258
Ashworth, A. 66

Bain, J.S. 20
Baker, M.R. & R.W. Cockfield 8
Barna, T. 9, 70
Buckley, P.J. 23, 48, 54
Buckley, P.J. & M.C. Casson 19, 25, 31, 33, 37, 47, 49, 50, 177, 179, 205, 264
  modal choice of market servicing 107-9
Buckley, P.J. & H. Davies 46, 60, 51, 106
Buckley, P.J. & P. Enderwick 193-4

Calvet, A.L. 54
Casson, M.C. 34, 37, 38, 41, 44, 50, 51, 54, 100, 113, 122-4, 177, 179, 193, 196, 225, 230
Casson, M.C. & G. Norman 30, 50, 100, 230

Caves, R.E. 21, 25-6, 87
Coase, R. 31, 32
Cockburn, C. 7
Cox, V.L. 8, 142

Demacopoulos, A. & F. Moavenzadeh 242
Dunn, A. & M. Knight 142, 242
Dunning, J.H. 16, 38, 47, 49, 51, 55-7, 85, 87, 177, 264
Dunning, J.H. & M. McQueen 86, 140
Dunning, J.H. & G. Norman 50, 86, 135
Dunning, J.H. & R.D. Pearce 16

Edmonds, G.A. 7
Eiteman, D.K. & A.I. Stonehill 233

Gabriel, P.P. 113
Graham, E.M. 29, 41, 50, 99, 225, 229, 264

Hamman, N.N. 8, 70, 71
Harrison, R.S. 62
Hawker Siddeley Group 8
Hennart, J.F. 40

Author Index

Hillebrandt, P.M. 67, 69, 73, 76
Hirsch, S. 21, 46, 49, 87
Hymer, S.H. 17, 18, 20, 51, 87, 131

Johanson, J. & J.E. Vahlne 205
Johnson, H.G. 21, 25, 87

Kaldor, N. 31, 32
Kim, S.H. 233
Kindleberger, C.P. 21, 87
Knickerbocker, F.T. 28, 41, 50, 99, 230
Kogut, B. 226-7, 229
Kojima, K. 51-4, 233-5

Lall, S. 22, 46, 101
Lessard, D.R. 40

Magee, S.P. 39, 48, 11
Mason, R.H., R.R. Miller & D.R. Weigel 22

Neo, R.B. 8, 14, 70, 83, 198
Norman, G. & J. Dunning 30

Ozawa, T. 52-4

Park, W.R. 68, 71
Pearce, J. 256
Penrose, E.T. 19, 31
Porter, M.E. 8

Robbins, S.M. & R.B. Stobaugh 234
Rugman, A.M. 38, 40, 47, 54

Stevens, G.V. 19
Stopford, J.M. & L.T. Wells 19
Swedenborg, B. 40

Teece, D.J. 44

Umeda, K. 146

Vernon, R. 26, 39, 40, 42, 48, 49, 50, 51, 225, 264

Westphal, L.E. 193
Williamson, O.E. 32-3, 35, 44, 54

Yannopoulos, G.N. 50